HOLLYW
U.S.A. ☆

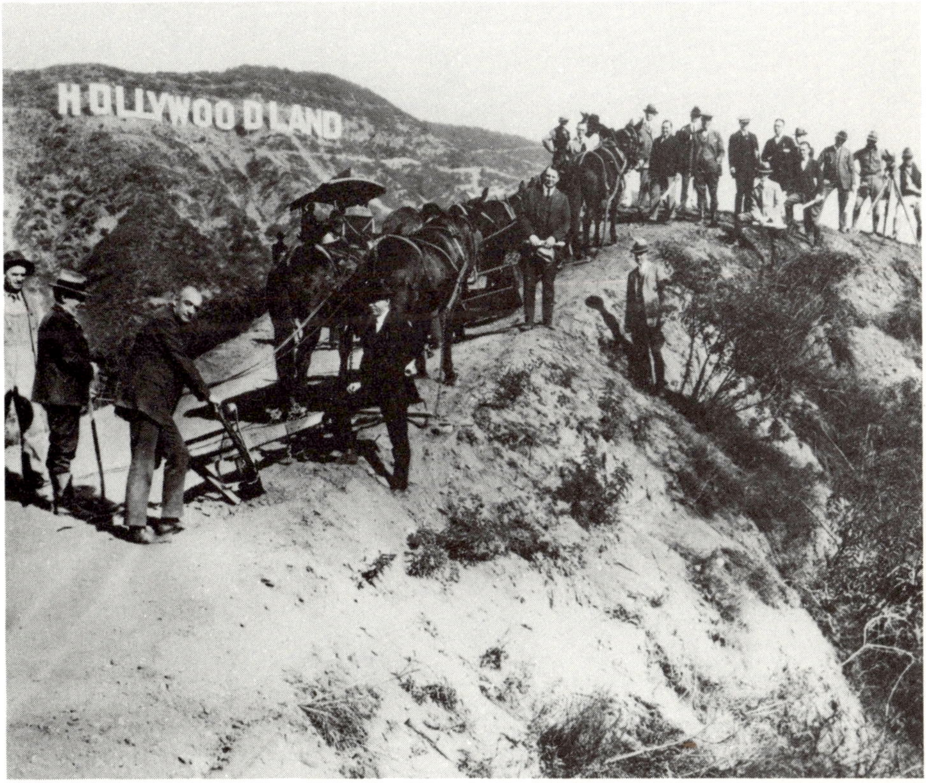

HOLLYWOODLAND U.S.A. ☆

The Moviegoer's Guide to Southern California

John Pashdag

CHRONICLE BOOKS
SAN FRANCISCO

Printed in the United States of America.

Library of Congress Cataloging in Publication Data

Pashdag, John.
Hollywoodland U.S.A.
Bibliography: p. 155
Includes index.
1. Moving-picture industry—California—Hollywood—History. 2. Hollywood Region (Calif.)—Description and travel—Tours. I. Title: II. Title: Hollywoodland USA.
PN1993.5.U65P37 1984 384'.8'0979494 84-4357
ISBN 0-87701-283-0

Photographic Credits:
Robert Landau: cover, pages 37, 38, 69, 70, 73, 77, 78, 84, 85, 89, 92, 100, 104, 107, 133, 145
Marc Wanamaker—Bison Archives: pages 10, 11, 58, 83, 86, 99
NBC Studios: page 21
Universal Studios: pages 14, 16, 138
Los Angeles Department of Arboreta and Botanic Gardens: page 55
Stan Adams: page 94

Book and cover design: Diana Fairbanks
Composition: Ann Flanagan Typography

10 9 8 7 6 5 4 3 2 1

Chronicle Books
870 Market Street
San Francisco, California 94102

Contents

Acknowledgments

Researching the history, or even the present state, of Hollywood is a task that even the most jaded sportswriter or copywriter would refer to as "awesome," but in the case of this book it was made easier by the assistance of a number of people. Chief among them was Marc Wanamaker, whose Bison Archives are an incredible source of both historical information and photographs of moviemaking in Los Angeles. Most of the historic photographs of Hollywood in this book come from his collection. Hollywood Heritage, the organization responsible for restoring the *Squaw Man* barn, was the source for much of the non-film-related historical information about the town of Hollywood.

Contributing writer Nancy Weldon was responsible for the sections on television-show tapings, game shows, screenings, cemeteries, hotels, and stargazing, while the section on memorabilia shops is the work of Jim Woller.

Finally, most of the modern-day photographs in the book are the work of Southern California photographer Robert Landau.

Introduction

When Don Gaspar de Portola led an expedition from Mexico to California's Monterey Bay in 1769, one of the priests accompanying him noted in his diary the existence of an Indian village at the mouth of a canyon that the Indians called "Cahuengna." Eighty years later, Kit Carson used Cahuenga Pass to deliver the mail from the United States to Monterey, and in 1874, the notorious bandit Tiburcio Vasquez was captured by a posse at Greek George's cabin just southwest of the pass.

In 1886, Kansas prohibitionist Harvey Wilcox bought 120 acres of the Cahuenga Valley and subdivided it. His wife, Daeida, named the new development Hollywood after the summer home of a woman she had met on a train.

And in 1911, the movies came to town.

There had been moviemakers in Los Angeles before. In 1898, Thomas A. Edison himself filmed a documentary short of downtown Los Angeles, and in 1903, a unit of the Selig Polyscope Company of Chicago filmed a ninety-five-foot, one-reel epic entitled *The Pigeon Farm at Los Angeles.* One can only feel empathy for those harried Chicagoans, who upon finishing their backbreaking twelve-hour workday, stopped off at the nickelodeon on their way home for a few minutes' solace watching Los Angeles pigeons.

A few years later, in 1907, Selig's Francis Boggs returned to California to film the climax of *The Count of Monte Cristo.* Boggs chose La Jolla for the dramatic scene of the count, having made good his escape, rising triumphantly out of the sea. The following year, Boggs rented the rooftop of a store in downtown Los Angeles, built a crude stage, and made *The Heart of a Race Track Tout,* the first full-length dramatic film shot entirely in Southern California. For his next epic, *The Power of the Sultan,* Boggs came down off the roof and rented a lot behind a Chinese laundry a few blocks away.

In the next three years a number of major film companies established bases in and around Los Angeles. In East Los Angeles were the Majestic

Film Co. and the Lubin Co.; in downtown were Biograph, with D. W. Griffith at the helm, and the New York Motion Picture Co.; out in Santa Monica were Kalem and Vitagraph. While some companies—those legally licensed by the Motion Picture Patents Company to use Edison's patented Kinetoscope equipment—were drawn by the sunshine and the wide-open locations, other companies—those that refused to pay the license fee, preferring to copy or steal their equipment—were drawn by the area's distance from the goon squads of the Patents Trust. California was still the Wild West, where an unlicensed director could easily get away with wearing a revolver on his hip, and if things got desperate, the Mexican border was just a hop, a skip, and a jump away.

One of those footloose independents was Englishman David Horsley, who arrived on the train from Bayonne, New Jersey, with a troupe of forty people in October 1911. Horsley's Nestor Film Company settled into the Blondeau Tavern, a wood-frame roadhouse at the northwest corner of Sunset Boulevard and Gower Street in Hollywood. The tavern had been shut down by local prohibition, and since the only thing Hollywood's founder's widow, Daeida Wilcox Beveridge, disliked more than drunkenness was movie people, the arrangement struck everyone except Mrs. Beveridge as the perfect match.

The new Nestor Studio had everything a film company needed in 1911. The corral housed the horses used in the Westerns, the barn held the props, twelve single-room structures became dressing rooms, and the main

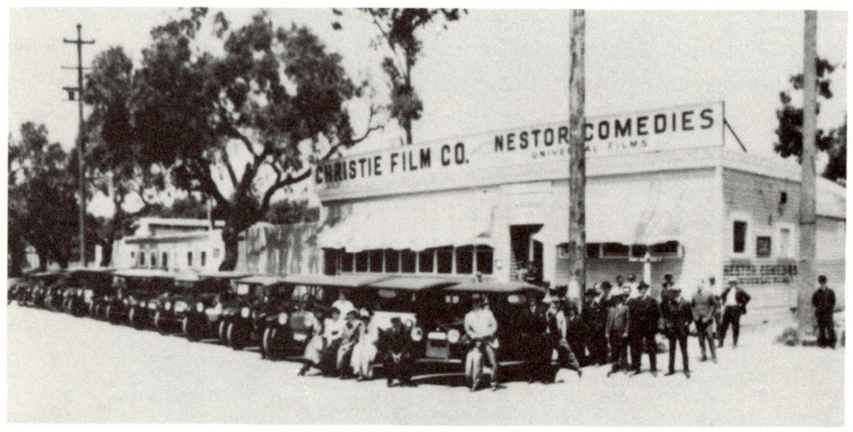

Hollywood's first motion picture studio, the Nestor Studio, moved into a former tavern at the corner of Sunset and Gower in 1911. By 1913, when this photo was taken, the movies were in Hollywood to stay.

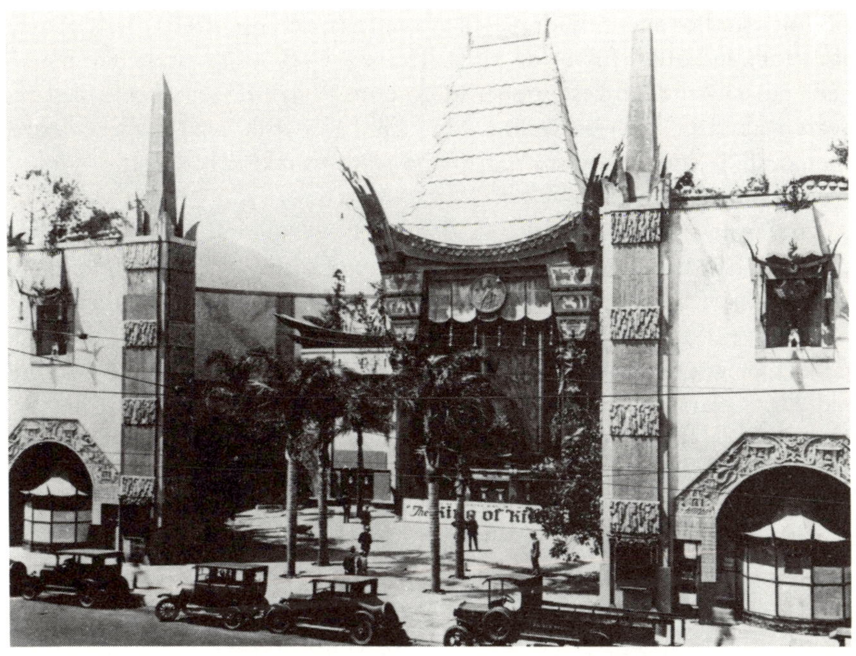

The Chinese Theater, built by showman Sid Grauman, opened in 1927 with the premiere of C. B. DeMille's King of Kings. *Hollywood's most famous tourist attraction looks every bit as spectacular today.*

part of the roadhouse, a five-room bungalow, was the main office. A forty-foot-square platform covered with sheets of muslin to diffuse the bright Southern California sunlight became the stage. The rental on the place was $35 a month, and Nestor made three films a week, a Western, a drama, and a comedy, for $400 apiece.

Within a matter of months, more than a dozen film companies were swarming in and around Hollywood. The mystique had moved in, and it's never left.

Today, of course, there's a lot more to Hollywood than just the town of Hollywood. The San Fernando Valley, Santa Monica, West Los Angeles, and especially Beverly Hills are all as much a part of the world of movies and television as is the intersection of Hollywood and Vine. Movies are made and stars live and dine all over town.

All of which makes seeing Hollywood difficult for those who don't know where to look.

Some of the sights of movieland, such as the Universal Studios Tour, the Chinese Theater, and Disneyland, are easy to find. But there are dozens of other studios and movie-related attractions, and hundreds of old movie locations and other historical sites. There are television-show tapings to see and star-studded restaurants to dine in. A half-dozen specialized research libraries cater to serious film scholars, while nearly three-dozen memorabilia shops cater to those who want to take a little bit of Hollywood home.

There's never been anything in history quite like Hollywood. No other place has had as much influence on the lives of so many people in so many different parts of the world at one time. The images that motion pictures have created over the last eighty-odd years have burrowed deep into the collective unconscious, creating new archetypes to substitute for the old. From Mary Pickford to Burt Reynolds, the Little Tramp to the Beaver, *Gone With the Wind* to *E.T.*, the movies have given us our biggest heroes, our strongest role models—and our most enduring dreams. To visit Hollywood is to visit more than just a town, or even, as the saying goes, a state of mind; whether you're from Los Angeles or London, New York or New Orleans, Paris or Peoria or Pickles Gap, going Hollywood can be just like going home.

I

HOLLYWOOD IN ACTION

The first thing most visitors to Hollywood want to see is a movie being made, and there are many ways to do just that. Two motion-picture studios and two television studios, one of which has a long history as a movie lot, give guided tours, and you can also watch film and television crews shoot around town on location. Most network television shows are taped in Los Angeles, and there's never any charge for tickets.

If you'd prefer to experience Hollywood as an active participant rather than simply as a spectator, you can try to get on a game show, a dance or exercise show, or one of the growing number of audience-involvement talk shows.

You can even try to snag a part in a genuine feature or television show. The odds are against you—way against you—but more unlikely things have happened.

Universal Studios

Is there anybody in the civilized world who doesn't know Bruce, the thirty-foot shark from *Jaws*? Bruce is just one of the stars of the **Universal Studios Tour,** by far the best-known studio tour in Hollywood.

Studio tours at Universal are an old tradition. When pioneer filmmaker Carl Laemmle bought a 230-acre chicken ranch in 1915 to house his Uni-

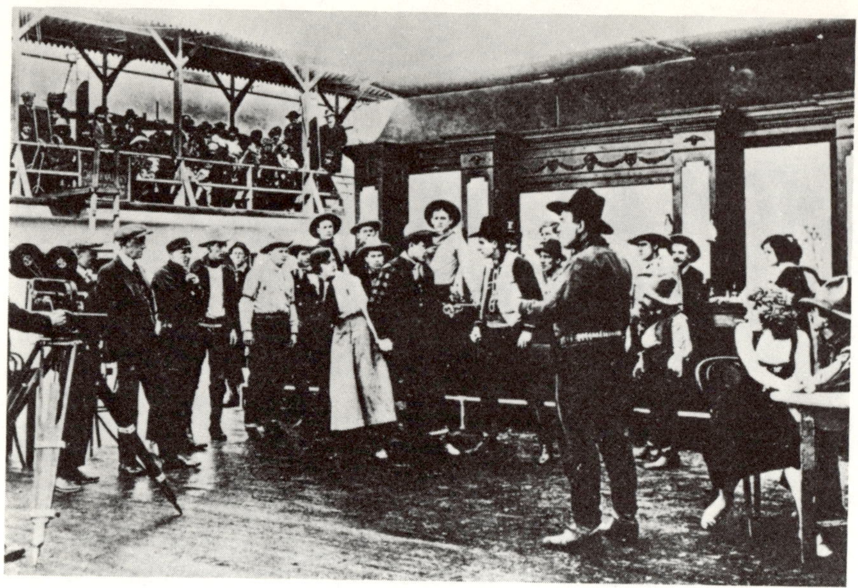

In the teens, "Uncle" Carl Laemmle charged a quarter to sit in these bleachers and watch his crews make Universal pictures. Today the tour costs a little more, but it's still a "must-see" for any visitor to Hollywood.

versal Film Manufacturing Company, one of his first actions was to open the lot to visitors, who could sit in a bleacher section for twenty-five cents and watch silent films being made. Since there was no soundtrack, the visitors were encouraged to cheer and boo to spark the actors' enthusiasm.

The coming of sound eliminated the studio visits for thirty-five years until 1964, when Universal's parent company, the Music Corporation of America (MCA), realized there was as much money to be made showing tourists how movies were made as there was actually making movies. Since then the tour has been expanded many times, so that in addition to the two-hour guided portion of the tour there are four free shows and a number of special exhibits, and visitors are advised to budget five or six hours to see the whole thing. The Tour Operations division employs more than thirteen hundred people full-time to service the nearly four-million annual visitors, and the tour's success has led MCA to begin construction of a new Universal City in Florida where filming will be just a side activity of the tour. Universal could go out of the movie business altogether and still rely on the studio tour to turn a buck.

The guided tour is still the main attraction, but during the peak periods of summer and spring and winter vacations you're almost forced to start the tour by seeing a couple of the free shows, whether you want to or not. When you enter the studio, you're given a tram boarding pass telling you to board the tram during a specified fifteen-minute period about an hour later.

The tram takes you through the front and back lots of the studio, with detours along the way to see a number of special exhibits and attractions. On the front lot, which is taken up with thirty-five sound stages, dressing rooms, and offices, the stops include the Costume Designers Galleries and the Special Effects Stage. In the former, a re-creation of Oscar-winning designer Edith Head's studio, you'll see costumes from new and old Universal releases ranging from *The Sting* to *The Best Little Whorehouse in Texas*, plus Head's eight Oscars and thirty-five nominations. In the latter, which was formerly Stage 30, you'll see—and volunteers from the audience will participate in—demonstrations of special effects ranging from the tilted-room effect used in the 1901 film *Heaven Help Us* to chromakey and rear-projection effects used in "Battlestar Galactica." Miniatures ranging from the pink submarine used in "Operation Petticoat" to the Ocean Park set from Stephen Spielberg's *1941* are on display, and in the chromakey demonstration, one member of the audience gets to don a "Hollywoodman" suit and pretend to fly.

The back-lot portion of the tour takes you through New York Street, used in everything from Abbott and Costello movies to "Quincy, M.E."; European Street, the location for *Frankenstein*, *All Quiet on the Western Front*, and television's "Rich Man, Poor Man"; and Six Points, Texas, an intersection of six Western streets dating back to the days of Tom Mix and Harry Carey, each with its own bar, bank, and jail. Six silent Westerns could be filmed simultaneously, and film crews could change locations just by swapping streets. Aspen Street, another Western street, has small doors on one side and tall doors on the other, so that short cowboys could be made to look big and tall dance-hall girls could be made to look petite.

The most popular sets, those that call up the most childhood memories, are Colonial Street, the one-time home of the Munsters, Marcus Welby, Bachelor Father, the Hardy Boys, and the Beaver, and, high on a hill overlooking the studio, the Gothic mansion used in *Psycho*. Most visitors have a hard time taking their eyes off the mansion as they pass, a lasting tribute to director Alfred Hitchcock's ability to create nightmares.

While the old sets are the most interesting part of the tram tour for movie buffs, Universal has added a number of special attractions in an attempt to broaden the tour's appeal, the most entertaining of which are the Battle of Galactica, in which the tram is captured by a Cylon armored vehicle and rescued by a Colonial warrior in a desperate laser battle; the *Jaws* shark attack, staged as the tram passes a replica of the New England town of Amity; the Doomed Glacier Expedition, a re-creation of an avalanche in an ice tunnel guaranteed to make the weak-stomached lose their lunch; and a flash flood through the back lot's Mexican Village. The other special effects—the collapsing bridge, the runaway train, the parting of the Red Sea—are, at best, second-rate.

Of the four live shows at the Visitors Center, the most entertaining is the Screen Test Comedy Theater. About two-dozen volunteers are chosen from the audience and, after a brief rehearsal, sent out on stage to re-create classic silent-comedy scenes, including a bank holdup, a chase on horseback, a leap into a raging river, some stunt flying, and a genuine cream-pie fight. The amateur actors are videotaped, the tape is edited with footage from real movies, and the resulting eight-minute epic is shown on television monitors. As you'd imagine, this is fun to watch but even more fun to take part in. If you've always wanted to be in the movies, this may be your best chance; get to the theater early and volunteer for every part. If you just want to watch, try to sit close to one of the monitors.

Universal's newest live show is The Adventures of Conan, which re-

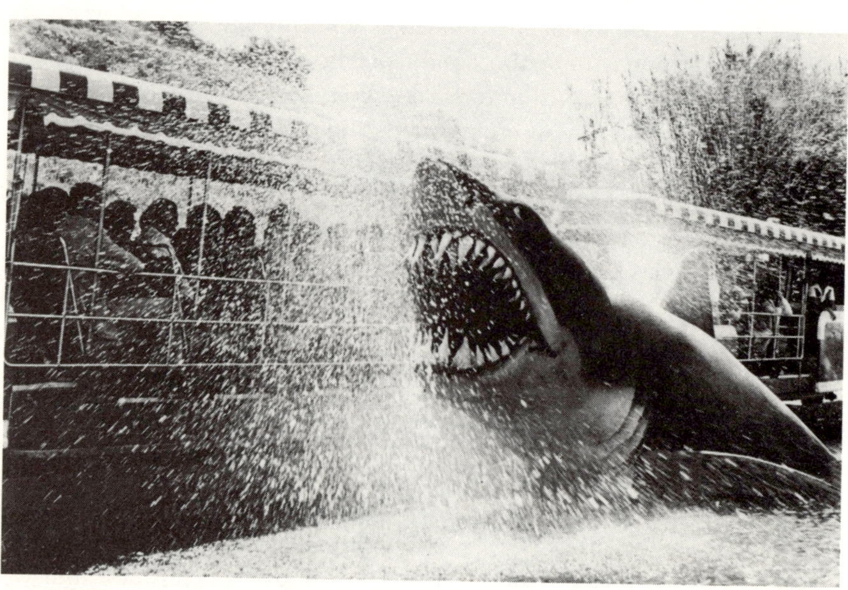

When you take the Universal Studios Tour, you can expect to meet at least one star close-up—Bruce, the mechanical shark from Jaws.

cently replaced the Castle Dracula. In this live re-creation of some of the spectacular effects from *Conan the Barbarian,* Conan and Red Sonja battle wizards and warriors, pitting swords against live lasers and fireballs, only to be confronted by an eighteen-foot-high fire-breathing dragon. Kids especially will love it. The two remaining shows, the live-action Western Stunt Stage and Animal Actors Theater, are old stalwarts well worth seeing for their demonstrations of the arts of stunt work, including fist fights, gunfights, and a twenty-foot-high fall, and animal training.

A must for animation fans is a visit to the new World of Woody Woodpecker in the Visitors Center, where the highlight is a fifteen-minute film retrospective of the evolution of Woody and the career of his creator, Walter Lantz. Exhibits include continuous cartoons on television monitors, a display of animation cels and related items, such as the first Oswald Rabbit model chart from 1928, and a large collection of Woody Woodpecker collectors' toys and other items. Kids can get free drawing lessons and can pick up a phone and talk to Woody in five languages.

Movie memorabilia of a more general nature is on display in the Motion Picture and Television Museum. There's a little of everything here. Start with a large collection of antique crank-operated motion-picture cameras and early moving-picture equipment, as well as the 35mm Mitchell camera used in filming *Gone With the Wind;* continue with a selection of props and costumes ranging from Errol Flynn's sword in *The Adventures of Don Juan* to one of Tom Selleck's "Magnum, P.I." Hawaiian shirts; move on to relics from all periods of the movies, such as Shirley Temple's teddy bear and Henry Fonda's Oscar for *On Golden Pond.* Take the time to look through the museum without rushing; it's not one of the most heavily promoted parts of the tour, but it's one of the most interesting.

A few words on how to survive five hours at Universal are in order. Try to arrive early in the morning and see the live shows before they get too crowded; the stunt show usually has the longest lines, so you might want to get it out of the way first.

If you go on a weekend or during a school-vacation period, be prepared to wait awhile for the tram. Once you get on, try to sit on the right-hand side for the best view.

Cameras are encouraged, and if you forget to bring a camera, you can borrow one for a returnable $20 deposit when you buy film.

Every Sunday, the guided portion of the tour is available in Spanish on a first-come, first-served basis. In addition, the tour is available in twenty other languages by reservation only; call 818-508-3771.

If you're traveling with a child, you can rent a stroller, or if with an invalid, a wheelchair, at the tour entrance. If your pet is accompanying you, you can put him up for free in a temperature-controlled kennel.

If you get hungry, try to avoid giving in to the junk-food booths scattered around the studio. Right outside the tour gate are three very good restaurants, Womphopper's Wagon Works, Fung Lum, and Victoria Station. The first serves barbecue lunches and dinners in a very stylish re-creation of an 1880s Western establishment; the second is the largest free-standing Chinese restaurant in the United States, built to resemble a classic Chinese palace; the third serves steak and prime rib in a fanciful reproduction of London's Victoria Station, complete with antique railroad cars once ridden in by Queen Victoria and Sir Arthur Conan Doyle and the departure board from the original London station. A cable car provides transportation from the station to the parking lot, and the tables in the old rail cars offer a spectacular view of the San Fernando Valley. You can also walk down the hill to the Sheraton-Universal's restaurant or coffee shop, either of which serves as many stars on a good day as the studio commissary.

The Universal Studios Tour is located on Lankershim Boulevard just off the Hollywood Freeway in Universal City and is open daily except Thanksgiving and Christmas. Summer hours are 8:00 A.M.–5:00 P.M. from mid-June to July and 8:00 A.M.–6:00 P.M. July and August. Winter hours are 10:00 A.M.–3:30 P.M. on weekdays and 9:30 A.M.–4:00 P.M. on weekends, with extended hours during winter vacation and other holiday periods. Admission is $10.95 for ages twelve and over, $7.95 for ages three to eleven, $8.45 for senior citizens, and free for children under three with an adult; parking is $1.50. For further information, call 213-877-2121.

The Burbank Studios

If you want to see the workings of a movie studio close up without having to put up with such distractions as laser-toting Cylons and runaway trains, **The Burbank Studios VIP Tour** is made for you. It costs a few dollars more than the Universal tour, it only takes about two hours instead of four or five, and you don't get to see a Western stunt show or an animal act. You will, however, get a much more personal look at a studio, since there are a maximum of twelve persons per tour, and, while the studio doesn't make any guarantees, there's about an 80 percent chance you'll be able to see real live stars making real television shows and movies, something you'll almost never see on the Universal tour.

The Burbank Studios was built as First National Studios in 1926. In 1928, the 108-acre lot was purchased by Warner Bros., and in 1972, Warner Bros. formed a joint holding company with Columbia Pictures, which bought the lot for the use of both companies. Besides Warner Bros. and Columbia, The Burbank Studios is now the home of about thirty independent production companies, including The Ladd Company, Rastar Productions, and Clint Eastwood's Malpaso Productions.

The VIP tour started as just that—a tour for visiting VIP guests of the studio. In 1973, the tours were opened to the public, although about 80 percent of the visitors are still guests of the studio since the tour does almost no advertising. While twelve is the largest tour group, there's no minimum number; if only one person shows up for the tour, that person will get a tour guide all to himself.

The tour is mostly a walking tour, and it's given rain or shine. If there are fewer than four people on the tour, it may be possible to take a golf cart, and occasionally the tour office may authorize the use of a small van if a real bigwig is along, but your best bet is to wear comfortable shoes and be prepared to walk.

The tour varies daily, depending on what's being shot and on weather conditions. The back lot is a staple, though. Streets include Tenement Street, built for *Annie* at a cost of $1 million; Laramie Street, used in "Bonanza" and "The Little House on the Prairie"; James Garner Street, home of "Maverick"; King's Row, built in 1941 for the film of the same name, which starred a B actor named Ronald Reagan, and used more recently as the seat of Hazzard County; New York Street, the oldest part of the back lot, which received its baptism of bullets in numerous James Cagney gangster movies; and the Jungle, formerly known as Walton's Flats.

The tour includes the inside of a sound stage where visitors may see the permanent sets from one of the many television shows filmed at The Burbank Studios, such as "The Dukes of Hazzard" or "Fantasy Island," and if there's shooting going on at the time of the tour, every attempt is made to take the visitors onto a working set where they can stand on the sidelines and watch the action and possibly chat with the stars between takes. Actually, in the twenty minutes or so that the tour visits a set, you're more likely to see eighteen minutes of preparation and a minute or two of actual filming, but that's still enough to tell your friends, "I saw them shoot that scene" when it appears on television.

Other stops on the tour might include the sound-recording studios, where visitors can see and hear anything from the McDonald's All-American High School Band scoring a commercial, to a rock group recording an album; the set-construction department, which uses enough material in one year to build 200 three-bedroom houses; or the prop department, which stores enough items to furnish 50 of those houses in any style from colonial to East Indian.

If you make reservations in advance, you can finish your tour by having lunch in the Blue Room, the executive dining room next to the studio commisary. The food, which ranges from salads and sandwiches to New York steaks, is very good and prices are very reasonable. If you stay long enough, you're bound to see some stars, and you can keep your table as long as you like. In addition, your waitress will tell you who has reserved a table that day and will usually act as an intermediary to get you an autograph if you ask.

When you make your tour reservation, the tour office will tell you if it's likely you'll be able to see actual filming. They can't guarantee anything, though, since they only get final clearance to visit a set ten minutes before a tour departs. Your *worst* chances to see actors in action are in March, April, and May when most television shows are on hiatus. Surprisingly, your *best* chances are on rainy days during the rest of the year, since more shows will be filming on the studio's stages instead of at nearby locations. Whether you see actual filming or not, though, The Burbank Studios VIP Tour is a much more authentic look at the real business of making movies than that of its much better-known competitor, Universal.

The tours are given at 10:00 A.M. and 2:00 P.M. Monday through Friday, with two additional tours at 10:30 A.M. and 2:30 P.M. in the summer. Admission is $18; children must be at least ten years old; and, due to union regulations, no cameras are permitted. The tour entrance is at the Hollywood Way gate off Olive Avenue in Burbank. Because of the very limited capacity of the tours, you *must* make reservations at least two or three days in advance by calling 818-954-1744 or 954-1008. You can make reservations by mail up to a month in advance if you send a check to confirm your space.

NBC Television Studios

Just down the street from The Burbank Studios is the West Coast home of the NBC television network and the **NBC Television Studios Tour.** This short tour gives visitors a very brief look at the workings of a major television studio, plus a glimpse at the most popular set in television land, "The Tonight Show Starring Johnny Carson." Sometimes, if the set is closed for rehearsal, the tour guide takes you out to the parking lot to show you Johnny Carson's car instead.

The forty-three-acre NBC Burbank lot opened in 1952 and became NBC's Los Angeles headquarters when the old NBC Studios at Sunset and Vine was demolished in the sixties. "You Bet Your Life" was filmed at NBC Burbank, as was "Rowan and Martin's Laugh-In" and most of the Bob Hope television specials. The tour is a one-hour walking tour with about two-dozen people in each group.

The first stop is a mini-studio with a control panel that the guide uses to show visitors how a real control booth works. At the same time he

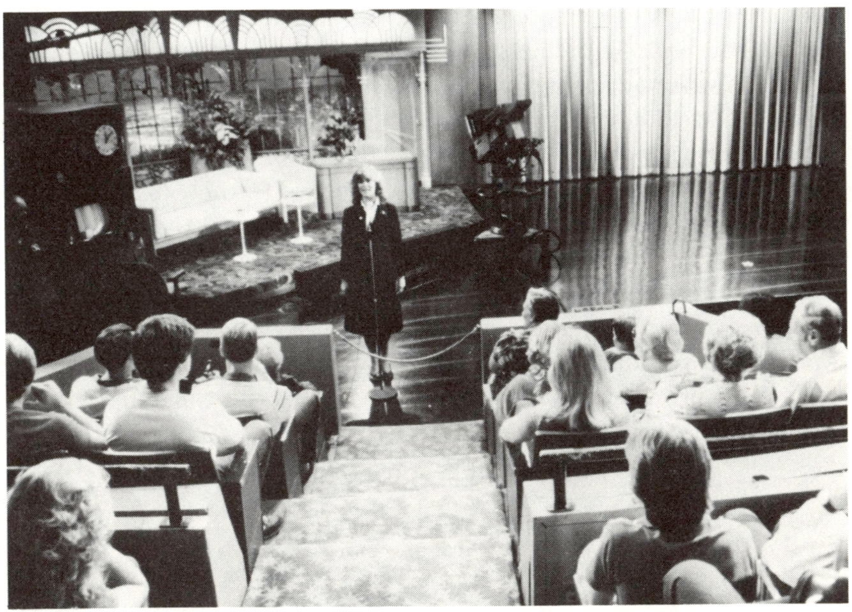

The NBC Studios Tour usually includes a stop at the set of "The Tonight Show," unless the set is closed for rehearsal; then the tour group goes out to the parking lot to see Johnny Carson's car.

demonstrates such techniques as wipes and dissolves and gives a step-by-step explanation of how a television show is produced. The explanation is very brief, but it gets high marks for at least attempting to explain the production sequence, something the motion-picture studio tours don't do.

Much of the tour is devoted to a verbal explanation of how the studio works, since there really isn't that much to see. Most of the sets used for different shows are rolled offstage when they're not in use, and there's nothing very exciting about visiting, say, the "Real People" stage when the "Real People" sets aren't there. "The Tonight Show" stage does interest Johnny Carson fans, but only the morning and early afternoon tours get to see it, since in the afternoon it's used for rehearsal. The other permanent set that most tours visit is the News 4 L.A. newsroom, where the guide explains such devices as the teleprompter and chromakey weather map, and reveals secrets like the fact that the impressive hi-tech switches and knobs over the newsmen's heads aren't hooked up to anything.

After a brief detour to the star parking lot to see—honest—Johnny Carson's car and the awning of the NBC Studios commissary, the tour continues at a rapid pace past (but not into) the set, wardrobe, and prop departments and the room where game-show prizes are stored.

I wonder who won the rickshaw?

If there's one major problem with the NBC tour, it's that you can easily spend longer waiting in line than on the tour itself. No reservations are taken, and when you arrive, you're assigned to a tour that may not leave for as long as an hour and a half during peak seasons. The earlier in the day you get there, the shorter the line. If you've got your heart set on seeing "The Tonight Show" stage, make sure you get there in time to take the 1:00 P.M. tour on weekdays, since the set closes at 2:00 P.M.

Tours leave every half hour 9:00 A.M.–4 P.M. daily except Thanksgiving, Christmas, and New Year's Day. Ages twelve and over pay $3.50, children under twelve pay $2.75, and senior citizens pay $2.50 weekdays and $2.00 weekends. NBC Studios is at 3000 West Alameda in Burbank; to find the tour office, turn south on California Street, at the west end of the studios. For further information, call 213-840-3572.

KCET Television Studios

When you take the **KCET Studios Tour** you're really taking two tours in one. On the one hand, you can see the totally modern television studio facilities of KCET-TV, Los Angeles's Public Broadcasting System affiliate; at the same time, you can see one of Hollywood's oldest existing motion-picture studios, with enough history behind it to fill a book.

Most of the buildings on the KCET lot date back to the twenties, but the studio has been in continuous operation since 1912, when the Lubin Manufacturing Company first put up a stage. The most important owner of the lot before KCET was Monogram-Allied Artists, which used it as a base for the East Side Kids and Charlie Chan, among others, while KCET now uses it for classier productions such as Carl Sagan's "Cosmos." The studio tour, which is a walking tour of about two hours, covers both the history and present use of the studio, with emphasis tailored for each tour group depending on its special interest. Groups are kept to a maximum of about a dozen people. The tour is easily the most educational of any of the Hollywood studio tours; visitors come away knowing something about every phase of television production.

After entering the old part of the studio through a brick gatehouse, the tour stops at Stage 3, now used for set design and carpentry, then moves on past a display of historical photos to the Hugh M. Hefner Little Theater, the oldest existing screening room in Hollywood. Built during the Charles Ray period in the twenties, the elaborate Spanish-style room, with its brick archways and concrete columns, had been boarded over and forgotten for nearly forty years until a KCET employee put his foot through the wall while demonstrating a karate kick and saw the brickwork underneath. Restored with a grant from Hefner, the room clearly shows what elaborate plans Ray had in mind for his company.

The tour includes the original Stage A, now used for "Why In the World" as well as local programs like "L.A. Week in Review," the studio control room, lighting booth, and audio room. Visitors are encouraged to sit at the various control boards, and the guide does an excellent job of explaining what everything is and how it works. A visit to Stage B sometimes includes a look at a rehearsal or taping of a new show, and while you're less likely to see major stars at work here than at The Burbank Studios, you'll at least get a glimpse of the real Hollywood at work.

All aspects of the post-production process are shown and explained on the tour, from editing, engineering, and sound recording, to master control, and since KCET is a busy place, you'll probably see nearly everything in action. Okay, so you can't see "The Tonight Show" set at KCET, but aside from that, the KCET Studios Tour has the NBC Studios Tour beat cold.

By the way, the KCET tour is absolutely free. Days and times for the tour change frequently. They're usually given at 10:30 A.M. Tuesday and Thursday, but if there's a group tour on a different day, you'll probably be able to join it. Children under ten aren't allowed, and you have to make reservations a day or two in advance by calling 213-666-6500. If you're more interested in the history of the studio than the present technology, or vice versa, you can request a guide with knowledge in that area.

The studio entrance is at 4401 Sunset Boulevard. After your tour is over, drive around to the back side of the studio on Sunset Drive to see how the exterior looked in the twenties.

Location Shooting

The first filmmakers who came to Hollywood did so for a number of reasons, one of which was the location. In addition to having clear weather most of the year, the Southland was the perfect place to find the Western locales needed for the cowboy shoot 'em ups that were so popular at the time. As filmgoers' tastes changed and Westerns fell out of favor, Los Angeles obliged by growing up and out in a variety of architectural styles, so that today a typical day's filming finds the city and the surrounding area doubling for everything from New York to Singapore, in addition to just being itself.

Visiting a location while shooting is taking place is one of the easiest ways to see moviemaking in action. There are no locked studio gates to talk your way through, and as long as you remain on public property and don't get in the way of the shot or the crew as they do their work, you'll be perfectly welcome to stay as long as you like. That may not be long; one of the first things you learn about filmmaking is that it's 5 percent getting the shot and 95 percent getting ready to get the shot, and that 95 percent is usually pretty boring to watch.

There are two ways to find out what productions are shooting where on any given day. The difficult way is to visit the offices of the Los Angeles City **Department of Public Works (DPW)** at City Hall, 200 North Spring Street (213-485-5328), which issues filming permits, and look through the permits on file. These are public records and there's no charge, but the permits don't tell you anything other than the location address and the name of the permit holder, which may not mean anything if you're not familiar with all the production companies in town. There are no film or television series' names, no stars' names, and only reluctant cooperation.

The easy way is to call **Hollywood ** On Location **** (8644 Wilshire Boulevard in Beverly Hills) at 213-659-9165. Each weekday morning, Hollywood ** On Location ** publishes a guide to that day's location shooting. The guide includes a list of 20 to 30 locations, including the name of the production; whether it's a feature, a television series episode or a movie for television; the stars of the production (who are *not* necessarily the stars on call that day); the address of the location and the hours it will be in use; whether it's an interior or an exterior; and whether there are any stunts scheduled. The guide also includes a map of the Southland showing where all the sites are located, plus detailed blow-up maps to lead you right to each location. Finally, you get an information packet with tips on how to conduct yourself while at the location.

Hollywood ** On Location ** sells a limited number of guides each day; you can reserve a guide by calling at least one day in advance and picking up your guide in person between 8:30 and 10:00 A.M. the following day. After 9:30 A.M., reservations are cancelled and remaining guides are sold on a first-come, first-served basis. Each guide costs $20.25 and must be paid for in cash.

Because of the increasing number of people visiting locations, the DPW recently tried to restrict access to permit information at the request of both homeowners concerned about the influx of tourists into their neighborhoods and production companies worried about interference with their work and damage to their already strained relationships with residents. To avoid becoming part of the problem, be as unobtrusive a visitor as possible. Park a block or two away from the site, don't stand in the middle of the street, and if a member of the crew or one of the policemen on duty asks you to move, then move. If the company is shooting interiors in a public place such as a hotel or an airport, you can usually enter, but if it's a private house or business, *do not* barge in. You can ask the assistant director's permission if you want, but don't expect to get a yes answer. Inside or out, stand off to the side, out of everyone's way, and when the director asks for quiet, be assured he means you, too.

If you *have* to try to get a star's autograph or photo, wait until he or she is on a break, and don't start cursing him out if he turns you down. Remember that the cast and crew are working, and just as you don't want anyone interrupting you when you're busy working, neither do they.

TV Tapings

Taking a studio tour isn't the only way to see Hollywood in action close up. Many television shows are taped or filmed in front of a live audience and, not only are tickets free, they're actually easy to get.

Most situation comedies that use a live audience instead of a laugh track are taped using the three-camera method pioneered in the 1950s by Desi Arnaz for "I Love Lucy." In a typical three-camera production both a dress rehearsal and a final performance are shot like a live play, sequentially, with a minimum of cuts or retakes. After the taping, the show's director picks the best scenes from both performances and edits them into the finished product.

In three-camera shows the audience plays as important a part as it would at a stage play; the actors hear a live response to their lines and often tailor their performances to suit. The laughs sound more real because they are, and the whole show picks up an air of spontaneity.

Of course, situation comedies aren't the only shows to use a live audience. Game shows, talk and variety productions, and dance shows rely heavily on, and actively recruit, real people to sit in the stands, enjoy, and react. And you can be a part of it.

Tickets are largely distributed in two ways: by the network or production companies themselves, and by professional headhunters, companies that are paid by the producers to round up all or part of an audience. Most situation comedies tape from mid-July through March; specials, pilots, and talk and quiz shows are shot year-round.

There are a few rules that apply to audience members. Most shows set an age limit, usually sixteen, but some will issue tickets to children as young as twelve. Leave your cameras and tape recorders at home; copyright regulations won't allow them in the studio. And plan ahead, especially for ticket requests by mail. It's simple, and there's no reason why you shouldn't be part of the show.

Live-audience shows broadcast on CBS are taped at seven studios in greater Hollywood. **CBS Television City,** 7800 Beverly Boulevard, Los Angeles 90036 (213-852-2624), can handle requests for all CBS shows no matter where they're shot. Send mail requests to the CBS Ticket Department giving the name of the show you want to see, your preferred date, and the number of tickets you'll need. Requests are filled in the order received and demand for tickets to some shows may limit availability. Actual tickets are mailed only within a 100-mile radius of CBS; people outside that area received guest cards that can be redeemed for tickets in person at the Television City Information Window. Tickets do not guarantee admittance, except for prearranged parties of twenty or more; call 213-852-2455 for group tickets. Tickets not distributed by mail can occasionally be picked up at CBS 9:00 A.M.–5:00 P.M. Monday through Friday and 10:00 A.M.–5:00 P.M. Saturday and Sunday.

NBC Broadcast Ticket Division, 300 West Alameda Avenue, Burbank 91523 (818-840-3537), handles ticket requests for shows taped at NBC, Paramount, and The Burbank Studios, from "Cheers" to "Fight Back with David Horowitz" but the hot ticket is always Johnny Carson, and some special rules apply to "The Tonight Show." To order by mail, send the name and desired date of the show and the number of tickets you want to "Tickets" at the above address. Orders are filled as received, but tickets are mailed only within a 150-mile radius of NBC; guest cards that must be redeemed for tickets at NBC on the day of the show are sent outside that area. Group reservations for twenty or more can be made by mail or phone (818-840-3555) and can be combined with a guided studio tour. A limited supply of tickets (first come, first served) can be picked up in person on the day of the taping at NBC and five other Los Angeles and Orange County locations (call NBC for information). And now, "Heeeeere's Johnny"; "The Tonight Show" limits tickets to four per request by mail and two per person at NBC on the day of the show. The age limit for the Carson show is sixteen. If you want to make sure Johnny himself will be the host, or want to find out the guest list, call 213-845-7000, extension 3061 up to three weeks in advance of the taping date.

While many of ABC's shows are taped at other studios, notably Paramount, tickets are arranged for all shows through the **ABC Television Center,** 4151 Prospect Avenue, Hollywood 90027 (213-557-7777). Mail orders should specify show, date, and number of seats, and include a self-addressed, stamped envelope. Allow four to six weeks for delivery; seats are *not* reserved, so be at the studio early. Limited tickets are sometimes available the day of the taping, but you must appear in person to pick them up. Special group arrangements can be made for most shows; call (213-557-4193 well in advance.

Dick Clark Productions, 3003 West Olive Street, Burbank 91523 (818-841-3003), provides tickets to the perennial teen favorite, "American Bandstand." Allow six to eight weeks for mail orders, which should specify a date and include a self-addressed, stamped envelope. Requests are restricted to those over twelve years of age. Call or write for schedules and group-order information.

Metromedia West Guest Relations, 5746 Sunset Boulevard, Hollywood 90028 (213-462-7111), handles ticket requests for the many syndicated shows and network situation comedies taped at the studios of Los Angeles station KTTV, a subsidiary of Metromedia. Minimum age for all shows is sixteen. Since most tapings here are by independent producers, a call or letter is recommended to determine schedules; tickets are available by mail order (allow four to six weeks) and occasionally in person.

Audience Associates, 1152 Gordon Street, Hollywood 90038 (213-464-8412), is a truly full-service independent ticket company. Normally, tickets for television shows only guarantee you a place in line; if the show is overbooked, you can be turned away. With Audience Associates, though, phone or mail queries (with a self-addressed, stamped envelope) bring *guaranteed* reserved seats at CBS, NBC, Metromedia, Universal, and KTLA. (ABC's policy is not to guarantee seats.) An Audience Associates bus also picks up audience members at the Chinese Theater on Hollywood Boulevard and will supply groups of forty or more with a free bus of their own within a one-hour radius of the studio. Should your group have its own bus, they'll give you some free gifts to raffle off on the way to the studio. You can get tickets the night before or sometimes the day of the taping, but no more than forty-five days in advance. If you're coming from out of town, wait until you arrive to make your request. Audience Associates will also arrange tours and restaurant reservations with your tickets, and after you've used their service once, they'll send you a periodic newsletter with information on upcoming shows.

Audiences Unlimited, 5746 Sunset Boulevard, Hollywood 90028 (213-856-1427), distributes tickets to shows taped at Universal Studios, including "The Jeffersons" and "One Day at a Time," as well as tapings at other Hollywood studios. Mail orders are handled in the order in which they are received and must be received at least three weeks prior to the taping. Visitors from out of state should send their requests six weeks prior to leaving for California. A self-addressed, stamped envelope is a must, and be sure to include the name, date, and time of the show you want to see. Write to Audiences Unlimited for a schedule of the shows offered. Two shows and six tickets per show are the maximum per request. You can also get tickets in person, on a first-come, first-served basis, for one week prior to each taping; the same two-show, six-ticket maximum applies. Groups of twenty or more can reserve tickets four to six weeks before the taping.

Representatives of **Television Audience Promotions,** P.O. Box 6968, Burbank 91510 (818-841-0111), can be found in the forecourt of the Chinese Theater on Hollywood Boulevard on most mornings handing out tickets to television tapings at ABC and KCOP studios. If dancing is your thing, you can join the crowd on "Solid Gold"; "Tic Tac Dough" and "Joker's Wild" will take care of game-show fans. Television Audience Promotions guarantees seats to many shows for out-of-the-area visitors (allow four to six weeks for mail order) and can arrange blocks of seats for groups who request them by phone. Mail requests must include a self-addressed, stamped envelope.

Game Shows

Admit it. Kissing "Family Feud" host Richard Dawson is your life's ambition, and you spend every waking hour hoping the neighbors never find out.

Or maybe you want to try your luck on "Wheel of Fortune"?

And you're quite positive you could come up with glib answers on "Tic Tac Dough" and win big.

In short, you know you could play the game as well as if not better than those other ordinary folks on television's game shows.

Well, as they say all too often in Southern California, "Go for it." But remember, the odds of getting on a show are low. Literally thousands of people call the various production companies every day. One program says that only about 10 percent of the people who apply make it on the air; others say the odds against getting on are much worse.

But, it's worth a try. So here are some general guidelines, plus specifics about contestant rules.

The classified section of the Sunday *Los Angeles Times* is a good starting point. Shows that need contestants run ads on the front page of the classified section; give them a call.

In general, most programs stipulate that contestants must be over eighteen years old and that they cannot belong to a performer's union. Often, contestants may not be related to production-company employees, and many shows disqualify people who have appeared on another game show within a year or on more than two other programs ever. In most cases, out-of-town contestants must pay their own travel and accommodation expenses. There are exceptions; we'll get to those.

Also, contestant coordinators say they want bright, outgoing, "real" people, so don't spend too much time in Hollywood or you will lose the "real" you. And don't forget that winnings are subject to taxation.

Goodson-Todman, 6430 Sunset Boulevard, Hollywood 90028 (213-464-4300), has several shows, each with slightly different requirements, but no written tests:

"Family Feud" staffers travel America searching out new families for nine months of the year, as well as interviewing extensively in Hollywood. Coordinators see more than eight hundred families a week, every year. Families audition; if they pass, they meet the producer who has final say on contestants. For out-of-town families, expenses to and in Los Angeles *are* paid.

"The Price is Right" contestants are chosen from the studio audience on the day of the taping. Call or write CBS Ticket Department for tickets (see TV Tapings). One hour before the show, the producer and assistants go down the audience waiting line, briefly (sometimes only a minute or so) chatting with everyone interested in being a contestant. The final choices are made, and Social Security numbers are checked to avoid sneaky repeaters.

"Tattletales" has celebs play the game. The studio audience is divided into color-coded groups, each of which is represented by a celebrity. If a group's celebrity wins, the group wins, too. Call CBS (see TV Tapings) for show tickets.

Merv Griffin Studios, 1541 North Vine Street, Hollywood 90028 (213-461-4701):

"Wheel of Fortune" invites telephone applicants (213-550-5555) to the office for an interview that consists of a puzzle-solving session. Hopefuls who pass are called in for a second, longer session of game-playing, which lasts about an hour and a half. If approved, the applicant has up to a year to appear on the show. Out-of-towners who return for the second session do so at their own expense, but then, they have time to perfect their skills.

The show also holds teen week, when contestants must be between thirteen and seventeen years old.

"Dance Fever" (213-461-4701) auditions dancers who win competitions held in clubs in thirty-five to forty cities each year. Once chosen, the lucky couples are flown to Los Angeles. They receive stereos, and may win up to $31,000 in cash and prizes with their fancy footwork. No professional dancers are allowed.

Barry & Enright Productions, 1888 Century Park East, Suite 204, Los Angeles 90067 (213-277-6000):

"Joker's Wild" contestants must pass a written ninety-question general-knowledge test, give in groups. Then, there's an interview, and after that, a series of interviews and run-throughs at the firm's Century City offices. Staffers try to expedite the requests of out-of-towners. They see some five hundred hopefuls a week throughout the year, but the odds may be worth it. One lucky contestant walked off with more than $300,000.

Reg Grundy Production Co., P.O. Box 67519, Los Angeles 90067 (213-557-3555):

"Sale of the Century" has a written general-knowledge test. If you pass, you return to play a mock-up of the game and be interviewed by staffers.

Bob Stewart Productions, 1717 North Highland, Los Angeles 90028 (213-461-3746):

"$25,000 Pyramid" tests prospective contestants on the first visit by having them play the game. After that, there's a second testing session. Contestants are usually called to appear within three months. Contact Bob Stewart Productions for information on how to become a contestant.

There's also a growing category of what might be called involvement or reality-participation shows:

"People's Court" is produced by Ralph Edwards Productions, 1717 North Highland, Hollywood 90028 (213-462-2212). Researchers go through small-claims court to find likely prospects. Participants must pass an interview session, and both sides must agree to abide by the judge's decision. The show pays the judgment, up to $500; participants split the remainder.

"Solid Gold" and **"Hour Magazine"** tape at KTLA-TV, 5800 Sunset Boulevard, Hollywood. If you'd like to watch a taping of "Hour Magazine," call 213-460-5256, or show up at the station one hour before the taping. To be in the "Solid Gold" audience, write P.O. Box 6968, Burbank 91510, and enclose a self-addressed, stamped envelope; or call 818-841-0110.

If you simply want to watch a game or talk show, call guest relations at the network on which the program is seen (see TV Tapings).

Casting

All right, so you've taken the studio tours, you've gone to watch filming on location, you've sat in the audience of a sit-com or "The Tonight Show" taping, but there's still one thing missing that would make your stay in Hollywood complete.

You want to be in a movie.

Well, lots of luck. Getting even a minor role in a movie requires so much hard work, talent, or luck, or all of the three, that you might as well try to win a sweepstakes.

Well, why not?

Every year, Universal Studios runs a "Win An Appearance" contest, the winners of which get to appear in Universal television shows. The contest is 99 percent a lottery and 1 percent a search for genuine talent. Contestants fill out entry blanks, available at the Universal Studios Tour Information Center around the beginning of December, and ten contestants are chosen at random. Those ten then take "screen tests" at the tour's screen-test theater (see Universal Studios), and their performances are judged by professionals from Universal's casting department. The number of winners varies annually, as each winner is given a bit part in a different Universal television show, and there's no way to predict how many shows the studio might have on the air in any given season. In 1983, five winners appeared on "Quincy, M.E." "Simon & Simon," "Tales of the Gold Monkey," "Knight Rider," and "Voyagers," and the very lucky grand-prize winner was flown to Hawaii to appear with Tom Selleck in "Magnum, P.I."

Even if you win, there's no guarantee that you'll actually be visible when the program is shown on television; a line of fine print on the entry blank gives Universal the right to edit out your performance, and yours could turn out to be the face on the cutting-room floor. Even so, if you don't have any of the aforementioned requirements for getting into the movies legitimately, Universal's contest may be as good a chance as you've got.

2

MUSEUMS AND ATTRACTIONS

At the mention of Hollywood and the entertainment business, most people think first of studios and stars and a town at odds with reality. Well, in many ways that's true, but Hollywood and its environs offer a number of institutions and attractions that can serve up a manageable slice of the dream factory and bring it all closer to home, a slice that you can reach out and touch, or at least see close up.

A good place to begin would be one of the Hollywood museums that have started springing up like wildflowers all around town. Here you'll find artifacts from every period and segment of the industry, from the silent age to the golden age to the television age. Hollywood's past is also chronicled and preserved at specialized libraries and archives, and a slew of specialty theaters, screening rooms, and film societies prove that there's more to Hollywood than just *Tootsie* and *E.T.* Finally, there are the major attractions such as Disneyland that wouldn't exist were it not for the movies and television.

Disneyland

Disneyland, America's best-known theme park, may have been spawned by the movies and television, but it doesn't have that much to do with them these days. True, many of the fifty-seven attractions in the six-acre park were inspired by Disney films—animated classics *Sleeping Beauty, Snow White and the Seven Dwarfs, Peter Pan, Alice in Wonderland, Dumbo, Pinocchio,* and *The Adventures of Ichabod and Mr. Toad,* as well as the live-action film, *Third Man on the Mountain*—and costumed actors representing world-renowned characters from the Disney pantheon wander throughout the park. In very few places in the Magic Kingdom, however, will you actually see movies themselves or items that were actually used in making movies. In fact, the only genuine movie prop in evidence is a steam locomotive used in the 1978 Don Knotts film, *Hot Lead, Cold Feet,* which sits on a siding below the mining town at the entrance to Big Thunder Mountain Railroad in Frontierland.

There are, however, plenty of items at Disneyland that have at least a secondhand connection to the movies, if only because so much of the park was built by men and women who had worked for the Disney Studios. In some ways, all of Disneyland is like one big movie set. The hippos, crocodiles, and other beasts that populate the Jungle Cruise in Adventureland, for example, were developed in part by Bob Wattey, who also worked on the giant squid for Disney's *20,000 Leagues Under the Sea* and on the mechanical shark in Universal's *Jaws,* and the painting of the 9.5 acres of rocks on the Big Thunder Mountain Railroad ride was supervised by Bob Jolley, whose previous sets graced the scenes of *South Pacific* and *The Greatest Story Ever Told.*

If you want to see movies at Disneyland, you have to go to Main Street, U.S.A. There you'll find the Main Street Cinema, which continuously shows five silent shorts, including Thomas Edison's 1903 *The Great Train Robbery,* the first film to have a plot, and, at eleven minutes, the longest film made up until that time. The cinema also shows one talkie, a cartoon called *Steamboat Willie,* in which a funny-talking rodent named Mickey made his debut.

Near the cinema is the Penny Arcade, where you'll find a number of Mutoscopes, hand-cranked silent-movie viewers built at the turn of the century. Developed by W. K. L. Dickson (who had previously developed the first motion-picture camera, the Kinetograph, and projector, the Kinetoscope, for Edison) for his own American Mutoscope and Biograph Co., these are the same kind of machines that were used in the original Nickelodeon and Biograph theaters in New York City. Instead of using film, each frame of the nineteen-second movies is printed on a separate card; as you turn the crank, the cards flip past the viewer. The machines here, which operate for a penny, show mostly Westerns and Charlie Chaplin comedies, with an occasional early animated short for good measure. Next to the Mutoscopes are a few Cail-O-Scopes, similar machines from about 1910.

Main Street is also the location of the Walt Disney Story, where Disney's offices are on display along with fifty-odd years' worth of Mickey Mouse memorabilia and other assorted Disneyana.

If you want to take home a souvenir with more class than a Mickey Mouse watch or Donald Duck key chain, stop in the Disneyana Shop, also on Main Street. Among the items for sale are original Disney animation cels, celluloid drawings used in the making of Disney animated films. Prices are roughly comparable to what you'd pay in collectors' shops in Hollywood, about $100 and up. Way up. The days when Disney animators considered these used cels junk and would jump into piles of them to slide across the floor are long gone. The shop also has a display of Disney collectables from years past; unfortunately, they're not for sale.

Disneyland's hours vary from season to season. In the summer, defined as mid-June to Labor Day, the park is open 9:00 A.M.–midnight Sunday through Friday and 9:00 A.M.–1:00 A.M. Saturday. From September to late March, hours are 10:00 A.M.–6:00 P.M. Wednesday through Friday, 9:00 A.M.–7:00 P.M. Saturday and Sunday, and closed Monday and Tuesday. From late March to mid-June, the times are 10:00 A.M.–6:00 P.M. Monday through Friday and 9:00 A.M.–7:00 P.M. weekends. In all seasons, hours are extended on holidays.

Crowds, like hours, vary from season to season as well. The most frantic times are the week after Christmas and most of the summer, while the lightest periods are the week before Thanksgiving, the week before Christmas, the first two weeks of June, and the last three weeks of the summer season. At any time, Saturday is the busiest day, while Friday and Sunday are the quietest.

Disneyland dropped its ticket books a number of years ago and now sells only "Passports," which include general admission and unlimited use of all rides and attractions, except shooting galleries and arcades. Prices are $13.00 for ages eighteen and over, $11.00 for ages twelve to seventeen, $9.00 for ages three to eleven, and free for children under three. Senior citizens (age sixty and up) pay $10.00 on Thursday and Friday. If you like to do things in an organized way, you can pay a few dollars extra for a 2½-hour guided tour, after which you're free to wander the park on your own. Parking is $1.00 extra.

If you're a photographer without a camera, you can borrow Polaroid still and movie cameras for free (after leaving deposits of $50 and $100 respectively) at the Polaroid Camera Center on Main Street. You can also rent Kodak Instamatic cameras for $2 and a $20 deposit, and there's a shop for minor repairs if your own camera breaks.

Kids or handicapped travelers in your group? The Stroller Shop at the main gate rents strollers and wheelchairs, the City Hall Information Center provides tape recorders and cassettes describing the park for blind visitors, and there are volume-control telephones located throughout the park for the hard of hearing. Dogs, cats, large reptiles? Only seeing-eye dogs are allowed in the park, but $2 a day buys a space in an air-conditioned kennel.

Disneyland is located at 1313 Harbor Boulevard in Anaheim, about twenty-seven miles southeast of the Los Angeles Civic Center. For information, call 213-626-8605 or 714-999-4565.

Walk of Fame

The **Walk of Fame** was created in 1958, when the Hollywood Chamber of Commerce decided to try to bring some glamour back to Hollywood by inscribing stars' names in coral terrazzo along the Hollywood Boulevard and Vine Street sidewalks. Each star has a brass phonograph record, radio mike, motion-picture camera, or television set at its center to show the field for which the star is being honored. By completion of the first phase of the project in 1961, there were 1,558 stars embedded in the sidewalks, with one per month being added since then for a current total of more than 1,700.

My own favorites are the four large stars at the Hollywood and Vine intersection honoring the Apollo moon landing astronauts, and the star in front of the Dan-Dee shoe repair shop on Vine Street honoring Maurice Diller. The star was supposed to honor Mauritz Stiller, the Swedish film director who discovered Greta Garbo, but by the time his star was placed in 1961, he had been pretty well forgotten—at least by the people in charge of placing the stars—and his name came out misspelled. Such is fame in Hollywood.

For years, the only way to find your favorite personality's star was to walk the entire length of the boulevard with your head down until you found it. Now the Hollywood Chamber of Commerce has published a list giving each star's name and address on the Walk of Fame. You can

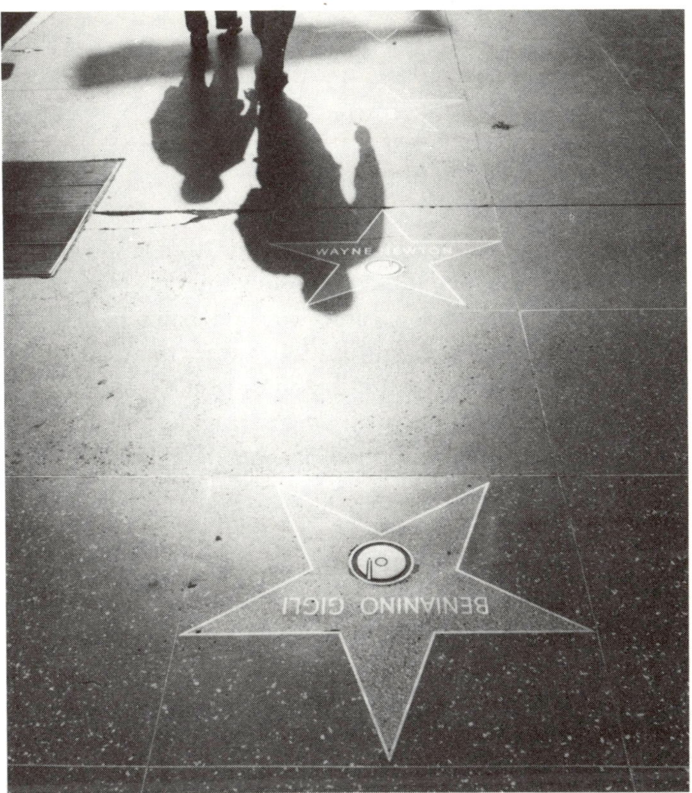

Hollywood Boulevard's Walk of Fame sometimes includes the not-so-famous.

buy the list for $5.00 at the C of C office located at 6290 Sunset Blvd., Suite 525, 9:00 A.M. to 5:30 P.M., Monday through Friday.

The Walk of Fame is bounded by Gower Street and LaBrea Avenue on the east and west and Franklin Avenue and Sunset Boulevard on the north and south. Once a month a new star is dedicated, with the star being honored usually in attendance. To find out who's being so honored, and where and when, call the Hollywood Chamber of Commerce at 213-469-8311.

Hollywood Studio Museum

The small, gaudily painted wooden barn that's now the home of the **Hollywood Studio Museum** is the oldest studio building left standing in Hollywood. Typically, not long ago it came within a hair's breadth of being demolished. Its move to its present location, across the street from the Hollywood Bowl, not only saved the building from the scrap pile, but also put an end to the more than twenty-year struggle to establish a motion-picture history museum in the heart of movieland, Hollywood.

The barn was built around 1895 on the Robert Northam farm at the corner of Selma and Vine streets in Hollywood, midway between Hollywood Boulevard (then Prospect Avenue) and Sunset Boulevard. In 1904, Jacob Stern bought the orange and lemon grove, and in 1913, Cecil B. DeMille and a troupe of players from the Jesse L. Lasky Feature Play

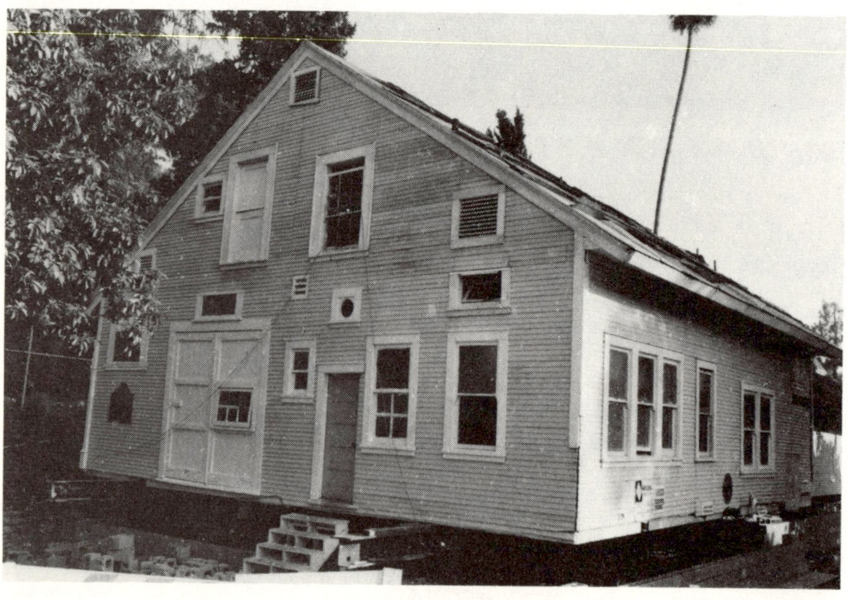

Hollywood's oldest existing studio building is the DeMille Barn, where Cecil B. DeMille and Oscar Apfel made The Squaw Man *in 1913. Even in those early days of the movies, extras lined up outside the studios looking for jobs. (1916)*

Company of New York arrived in Los Angeles, looking for a location to shoot their version of *The Squaw Man*, a play which had been a hit in New York. With them was their star, Dustin Farnum, who had had a supporting role in the New York cast, and their director, Oscar Apfel. DeMille rented the barn and part of the orange grove for twenty-five dollars a month, after getting approval from Lasky's vice-president, Sam Goldfish, later Sam Goldwyn of Metro-Goldwyn-Mayer (MGM).

Part of the barn was still being used for horses, and when the barn was washed down, DeMille would put his feet in a wastebasket next to his desk to keep them dry.

The Squaw Man started shooting on 27 December 1913, less than two weeks after DeMille moved into the barn, with the experienced Apfel at the helm and DeMille, who had never directed a film, assisting. The total cost of the film was $15,000, and when it grossed more than fifteen times that amount, Lasky's company quickly took over the whole block.

While Lasky was building his Feature Play Company, another producer named Adolph Zukor was building up his Famous Players Film Company. In 1916, Zukor and Lasky merged to create the Famous Players-Lasky Corporation. A decade later, when the combined company bought out their distributor, they purchased another studio at Marathon and Van Ness, moved the entire company to their new lot, and eventually took the name Paramount.

The old studio was demolished except for the barn, which went along to the new studio, where it became dressing rooms, a library, a storage building, and eventually the studio gymnasium. In 1956, it was moved again, this time to Paramount's Western Street, where it became the railroad station in the television series "Bonanza." That same year, it was declared a California State Landmark.

By 1979, Paramount's back lot was needed for other things, and the barn where Paramount had started sixty-six years before was old and in the way. It was offered to the Hollywood Chamber of Commerce, who moved it to a parking lot next to the Palace Theater on Vine Street, where it continued to deteriorate. Finally, in 1982, the barn's wandering came to an end when it was turned over to the Hollywood Heritage committee and moved to its present site, originally acquired for a Hollywood museum twenty years ago, where restoration began in earnest.

Current plans call for the museum to be open by spring of 1984, with a nominal admission charge. Because of the small size of the barn, the collection will be limited to relics of the pioneer days of motion pictures, with exhibits showing the development of the studios, and old props and equipment, including the camera used in the filming of *The Squaw Man.* A small screening room will be devoted to silent pictures. That may not sound like much, but considering that the barn itself is at least as historic as anything that could be displayed in it, it's definitely worth seeing nonetheless.

The Hollywood Studio Museum is at 2112 Las Palmas Avenue, just off Highland Avenue in the parking lot across from the Hollywood Bowl. For more information, call 213-874-BARN.

Libraries

Most people know the **Academy of Motion Picture Arts and Sciences** as the folks who hand out the Oscars every year, but there's more to it than that. The Academy's headquarters complex at 8949 Wilshire Boulevard in Beverly Hills (213-278-8990) houses a number of operations of interest to every movie fan from the serious researcher to the merely starstruck. The goal of the **Margaret Herrick Library** (213-278-4313) is to collect every book published in English on motion pictures; currently they have over 14,500 volumes dating from 1912. Besides 200 current periodical subscriptions, the library has a large collection of trade publications dating as far back as 1906. There are production files on over 93,000 movies—virtually every American film ever made, as well as major foreign works. Special collections include 500 scrapbooks detailing the careers of Hollywood notables, stills and scripts from 2,200 Paramount films, the Mack Sennett and Selig Collections, and the still collections of Thomas H. Ince, RKO Radio Pictures, and MGM. The Herrick's material is noncirculating; photocopies cost fifty cents and copies of stills are ten dollars.

Downstairs at the Academy's showcase **Samuel Goldwyn Theater,** there's more to see than film retrospectives and specials (see Screenings). The foyer of the Goldwyn features a permanent display of set sketches and posters from Oscar-winning films. In the main lobby below, you can see revolving displays of a wide range of other film-related material from star portrait photographs to special-effects matte paintings, and a display case outside the library features revolving exhibits of memorabilia relating to film personalities from John Ford to James Dean. The Academy exhibit areas are open 9:00 A.M.–5:00 P.M. Monday through Friday; the

Herrick Library is open 9:00 A.M.–5:00 P.M. Monday, Tuesday, Thursday, and Friday.

The **American Film Institute** (AFI) at 2021 North Western Avenue (213-856-7600), is the home of the **Louis B. Mayer Library,** a reading, reaference, and research facility designed for the faculty, fellows, and staff of AFI's Center for Advanced Film Studies, but open to the public as well. The collection has more than 5,000 books on all aspects of film, television, and video, as well as works on the theater. More than 2,250 unpublished feature-film shooting scripts and 800 television scripts are available, as are 150 current periodicals, an entertainment clipping file, transcripts of the AFI seminars and oral history projects, a number of special collections of personal papers and manuscripts of leading film figures, and the Columbia Stills Collection, covering the years 1930–50. All materials are noncirculating. The library is open 10:30 A.M.–5:30 P.M. Monday through Friday. AFI also sponsors numerous screenings (see Screenings) and seminars, many of which are open to the public; call for information.

While the staff at the **Walt Disney Studio Archives,** 500 South Buena Vista Drive, Burbank (818-840-5424), are often asked to settle trivia contests and are happy to oblige, their huge collection of Disney-related material is generally available only to industry members and serious researchers. Those who qualify can examine over eighty-five hundred photos of Walt Disney, scripts, production files, advertising and promotion materials, books, and even oral histories taken from the folks who helped build the Magic Kingdom. There are complete files, including engineering and architectural material, on Disneyland, Disney World, and the new Tokyo Disneyland. The collection of Disney character merchandise is the most complete in the world, and the staff can offer opinions on the authenticity of your Disneyana and give recent auction prices, although they will not appraise items. Researchers' inquiries should indicate their area of interest; the archives will send a bibliography and suggest materials that can help. The archives are open 9:00 A.M.–5:00 P.M. Monday through Friday, by appointment only.

The **University of California at Los Angeles** (UCLA) **Theater Arts Library** in the University Research Library on the campus in Westwood (213-825-4880) recently added the Charlton Heston Collection to its already enormous hoard of film material. Now you can see the staff Moses turned into a snake in *The Ten Commandments,* along with career memorabilia from more than two hundred actors, writers, and directors. Revolving exhibits are displayed in the library's main lobby and in the foyer outside the Theater Arts Library. The library is open to the public; persons interested in the special collections (like the Heston items) should call or write ahead for an appointment, but the thousands of volumes of theater and film lore are available at the desk. The library is open 9:00 A.M.–11:00 P.M. Monday through Thursday, 9:00 A.M.–6:00 P.M. Friday, 9:00 A.M.–5:00 P.M. Saturday, and 1:00 P.M.–10:00 P.M. Sunday, with restricted hours during school vacations; special collections can be examined 9:00 A.M.–4:00 P.M. Monday through Friday.

The **University of Southern California (USC) Cinema Library** in the University Library on the campus near downtown Los Angeles (213-743-6058) includes the Warner Bros. Studio collection and the partial Universal Studios collection, featuring scripts, production files, story files, legal work, stills, and studio histories. In addition, the library has more than 175 special collections, covering the careers of industry leaders from Jerry Lewis and Irene Dunne to Cecil B. DeMille and Hal Roach. While this material is not on display, the Cinema Library is open to the public 8:30 A.M.–5:00 P.M. Monday through Friday and 9:00 A.M.–5:00 P.M. Saturday. Write or call ahead so the staff can locate the material you need.

Movie Memorabilia

The most incredible collection of science fiction and fantasy film memorabilia anywhere is the **Forrest J Ackerman Collection,** open to the public on Saturday afternoons by appointment only. Ackerman, who was editor of *Famous Monsters of Filmland* magazine for 25 years, has more than 300,000 items filling every available square inch of his house in the Hollywood Hills, including the garage, backyard, and basement. In addition to 125,000 stills and 18,000 lobby cards, the king of fantasy-film fandom is the extremely proud owner of such one-of-a-kind items as the cape Bela Lugosi wore in *Dracula,* the full-head masks worn by the Metaluna Mutant in *This Island Earth* and by the Creature from *The Creature from the Black Lagoon,* and the body mold that Rod Steiger wore to become *The Illustrated Man.* His vast collection of props includes a Martian war machine from *The War of the Worlds.* Ackerman spent fifty-seven years acquiring his collection and has offered to give it all to the city of Los Angeles if the city can find a suitable location for it. For now, the public is invited most Saturday afternoons from 2:00 to 4:00 P.M. You must call 213-666-6326 the day before to make arrangements and get the address.

It may not actually be in Hollywood, but there's more movie memorabilia on display at the **Variety Arts Center,** 940 South Figueroa Street in downtown Los Angeles (213-623-9100), than on any six blocks of Sunset. The center is a five-story city cultural monument and headquarters of the nonprofit Society for the Preservation of Variety Arts. It's open to the public for lunch (Monday through Friday 11:30 A.M.–2:30 P.M.), dinner (6:30 P.M.–10:30 P.M.), and performances in three different theaters, and there's no charge for looking.

In the W. C. Fields Bar you can see the largest collection of Fieldsiana in the world; the trick pool table and bent pool cue he used on stage and in film are in an adjacent library. The Earl Carroll Lounge features the autographed cement blocks from the walls of that showman's Hollywood theater, while on the theater Roof Garden, you can sip your favorite drink in the restored Harmonia Gardens bar used in *Hello, Dolly.* Every room is crammed with theatrical and film memorabilia, playbills, and props, and three libraries, one for film-related materials, one for variety-arts material including gag files and burlesque sketches, and one for sheet music, preserve the heritage of show business. The music library includes a one-of-a-kind collection of silent-movie sheet music, most with self-explanatory titles like "Cliffhanger" and "Love Theme for General Use." Call the center to check on the free Wednesday noontime concerts (brown baggers welcome) and the various live productions offered.

The **Hollywood Wax Museum** is a wax museum of the traditional variety—slightly seedy, slightly ragged, with wax figures that are often downright unrecognizable, and sets that have seen better days. It is redeemed only by a Chamber of Horrors that is always gruesome and can be genuinely spooky under the right conditions. Started fifteen years ago by Indian-born Spoony Singh, the museum has more than 220 figures on display. Several presidents, historical figures such as Amelia Earhart and Joseph Stalin, and a re-creation of the Last Supper share the cramped, dingy museum space with the usual movie and television stars. Most look more like distant relatives of the people they model, and some are identifiable only by their name plaques. The Old Palace Theater within the museum, a trashy, broken-down room with steps for seats, presents a continuous screening of stills from Academy Award–winning films interspersed with occasional film clips of winners accepting their awards. The phrase "third-rate" takes on a new meaning here.

The Chamber of Horrors, on the other hand, isn't bad. About half the figures in this section are old familiars like Dracula, the Mummy, and the Wolf Man, while the rest are such horror chamber regulars as medieval torturers and their victims. The number and explicitness of the torture scenes make the room a little too grisly for young kids, but high schoolers can—and do—scare themselves silly in a place like this.

The Hollywood Wax Museum is at 6767 Hollywood Boulevard (213-462-8860). It's open 10:00 A.M.–midnight Sunday through Thursday and 10:00 A.M.–2:00 A.M. Friday and Saturday. Admission is $5 for ages eighteen and over, $3.50 for ages thirteen to seventeen, senior citizens and armed forces personnel in uniform, $2 for ages six to twelve, and free for children under six with an adult.

Six Flags' Movieland in Buena Park, a block from Knotts' Berry Farm, is the larger and more elaborate of the Southland's two wax museums, but don't pass it by just because you don't like wax. Surprisingly, it also turns out to have one of the best collections of early silent-movie shorts and nickelodeon-style viewers.

The wax museum itself is unexceptional. More than two-hundred wax statues of stars from Shirley Temple to William Shatner are displayed in re-creations of sets from their most famous films and television shows, and a number of the figures are dressed in original costumes donated by the stars themselves. Even though the sets and the costumes are reproduced fairly authentically, that doesn't make up for the fact that you're still just looking at wax. While the wax figures do, in some cases, look surprisingly lifelike, most of them are not very accurate reproductions of the well-known faces. It's almost as if they used look-alikes for their models instead of the real stars.

Movieland has been trying to expand on its wax museum image recently by opening new attractions, the most ballyhooed of which is the Black Box, a re-creation of sets from *Halloween, Altered States,* and *Alien,* complete with sound and lighting effects and wax figures that pop out of closets and such. If Movieland really spent $1 million on the Black Box as advertised, they were gypped. The only set that's at all scary is the one from *Alien,* and that's just because of. . . well, why spoil it?

Two other new attractions are also poorly done. Backlot Boulevard is a very short, narrow aisle fronted by sets purchased from Laird International Studios in Culver City, where they were once used in the films *Citizen Kane* and *Gone With the Wind,* but there aren't even photos posted to show how they looked in those films, and the overall effect is disappointing. Starprint Plaza outside Movieland tries to continue the Chinese Theater's tradition of recording stars' footprints and autographs in cement, but it's hard to get excited when the biggest stars in the plaza are Vincent Price, Herve Villechaize, and Los Angeles television horror-film hostess Elvira.

There's one attraction, though, that makes Movieland definitely worth seeing for motion-picture history students and silent-film buffs: the world's largest collection of Mutoscopes, similar to the ones displayed at Disneyland but much more numerous and much more varied in subject matter. The Mutoscopes are scattered throughout Movieland and are grouped roughly by genre. The first batch has about twenty early nudies—more daring than you'd expect for 1905—with enticing titles like *The Doctor's Office, After the Bath,* and *Eve's Leaves.* Other subjects throughout the museum range from early Chaplin comedies and Westerns, to Felix the Cat and other cartoons, and nature studies like *Killing the Killer,* a fight between a mongoose and a cobra. Most of the machines cost twenty-five cents to operate (up from the original nickel), a few cost ten cents, and there are dollar-bill changers in some locations. You have to turn the crank a few times before the picture starts. My advice on visiting Movieland if you're a silent-movie fan is to forget about the Black Box and the wax figures, but be sure to carry a large supply of dollar bills or quarters.

Movieland is at 7111 Beach Boulevard in Buena Park (213-583-8025 or 714-522-1154). The museum and all attractions are open 9:00 A.M.–9:00 P.M. Sunday through Thursday and 9:00 A.M.–10:00 P.M. Friday and Saturday during the summer, and 10:00 A.M.–8:00 P.M. Sunday through Thursday and 10:00 A.M.–10:00 P.M. Friday and Saturday during the winter. Admission is $7.50 for ages twelve and over, $4.50 for ages four to eleven, and free for children three and under. A ticket stub and $3 buys an annual pass good for unlimited visits.

In the early sixties, veteran Hollywood stunt pilots Paul Mantz and Frank Tallman combined their collections of rare, antique, and stunt airplanes in the **Movieland of the Air** museum at Orange County's John Wayne Airport, 19711 South Airport Way in Santa Ana (714-545-1193). Mantz and Tallman are gone, killed in separate plane crashes (Mantz while filming *Flight of the Phoenix*), but the collection still survives. Amid reproductions of movie sets rest a World War I Fokker triplane and an arch rival Sopwith Camel used in *The Great Waldo Pepper,* among other films. Two B-25 Mitchell bombers from *Catch-22* and World War II fighters flown in hundreds of movie dogfights can be seen in person and in stills and posters of their starring roles. The museum is open 10:00 A.M.–5:00 P.M. Wednesday through Sunday seven days a week during the summer. Admission is $3.00 for adults, and $1.00 for children under twelve.

Other Museums

Besides the museums that specialize in motion-picture memorabilia, there are three other Hollywood-area museums that at least occasionally have film-related exhibits.

The **Craft and Folk Art Museum** at 5814 Wilshire Boulevard (213-937-5544) across from the Los Angeles County Museum of Art, is one. Most of the museum's shows are in traditional categories, such as Mexican village folk art or modern Finnish design, but one of the most popular shows of recent years was "The Art of the Dark Crystal," a 1982 display of the puppets used in the making of that remarkable film by the team that created the Muppets.

The **Junior Arts Center Gallery** in Barnsdall Park at 4800 Hollywood Boulevard (213-666-1093)„ is another museum that sometimes gets into the area of film. Its 1983 show, "Make Believers: Movie Special Effects," rivaled the Universal Studios Tours' Special Effects Stage as an entertaining introduction to the art of movie magic. Since the Junior Arts Center's main business is teaching children, most exhibits include unique classes and workshops aimed at younger visitors, but of interest to all ages. How often, for example, do you get a chance to make your own special-effects movie clip on videotape, using miniatures and matte paintings from *The Sting, Treasure Island,* and *The Black Hole?*

The **California Museum of Science and Industry** at 700 State Drive in Exposition Park (213-744-7400) occasionally features special exhibits relating to the technological aspects of moviemaking. Fans of *E.T.*, for example, could have seen the evolution of their hero from sketchbook to film in the recent show about E.T.'s creator, "The Man Who Made the Film Creatures: Carlo Rambaldi." The museum is open 10:00 A.M.–5:00 P.M. daily.

Screenings

Contrary to popular belief, sex, drugs, and financial flimflam are not the raison d'être of Hollywood. Movies are.

Moviemakers are close at hand, available to attend seminars and retrospectives on everything from women in cartoons to B-movie monsters.

Special screenings abound. Always call ahead to check on location changes and ticket availability.

Topping the list of screening facilities is the **Academy of Motion Picture Arts and Sciences,** 8949 Wilshire Boulevard, Beverly Hills 90211 (213-278-8990). The Academy, home of the Oscars, offers three or four film events a month, open to members and to the public by fee. For no more than the price of a regular movie ticket, film buffs can attend retrospectives of the works of industry greats of the past and present. Programs usually include film clips and a panel discussion with actors and filmmakers. Retrospectives on stars like Mae West, Clara Bow, and Judy Garland often sell out far in advance. The Academy Foundation at the above address offers a mailing list of events for just three dollars a year. And remember, there are usually no public screenings in February and March, when the Academy is busy showing members films nominated for Oscars.

Most of the films screened at the prestigious **American Film Institute (AFI)**, 2021 North Western Avenue in Hollywood (213-856-7600), are for the institute's students and guests only. A few special film programs are open to the public by fee, however, and AFI does sponsor various seminars, on topics from screen writing to international marketing, which are also open to the public by fee. Meanwhile, contemporary video artists are now showcased in the "Screening Room West" series. Videotapes, ranging in content from music and satire to documentary, are shown in the Mark Goodson screening room at the AFI campus.

The **Leo S. Bing Theater** in the film department of the **Los Angeles County Museum of Art,** 5905 Wilshire Boulevard, Los Angeles (213-857-6201), offers a wide range of foreign films and star tributes throughout the year. Most showings are on Friday and Saturday nights (except during the Christmas and New Year holidays when the theater is closed). The admission fee is low, with discounts for students and museum and AFI members. The Bing is a comfortable theater. Serious moviegoers will be glad to know that talking patrons and howling babies are frowned upon here. Tickets often sell out, so call in advance.

The **Los Angeles Film Exposition,** better known as **Filmex** (6525 Sunset Boulevard), is an annual three-week spring event that offers nearly two hundred feature films and shorts. Filmex brings out the serious and trendy film followers, as well as people who know somebody who worked on a particular film. Almost everybody sits through all the credits, which can include such obscure mentions as "sister to the director." The festival shows foreign films and minor or independent American ones that have never had a commercial release. Good fun, and plenty of choices for viewing. Call 213-469-9000 for information; screening locations may change from year to year.

There are many other somewhat-lesser-known screening groups. Both the University of California at Los Angeles (UCLA) and rival University of Southern California (USC) have screenings, which are open to the public.

Melnitz Hall at **UCLA** (213-825-2581) has one of the best deals in town. The films are almost always free, and screenings are held most nights during the school year and also during the summer. There's a little bit of everything, from animation, mysteries, and classics, to gay-oriented works, and some tributes and seminars, too.

DKA, the student film society of **USC**, offers special screenings, series, and retrospectives on weekends throughout the school year and some summer programs, all at the Norris Cinema Theater (213-747-0783). The events are usually free to the public.

Then, there are special interest groups:

The **Academy of Science Fiction, Fantasy and Horror Films,** 334 West 54th Street in Los Angeles (818-842-1351 or 213-752-5811), screens some seventy films a year. The shows are open to members for a forty-dollar annual fee, or to out-of-towners for a ten-dollar fee. Also, some free public seminars on acting, special effects, and other aspects of filmmaking are held.

The **American Society of Amateur Filmmakers,** 1200 Riverside Drive, No. 165, Burbank (818-842-1351), has free screenings of super 8 and 16mm films by amateurs who ask viewers to fill out critique sheets. Celebrity guests who started out as amateurs sometimes appear as moderators.

The **Black American Cinema Society,** 3617 Montclair Street, Los Angeles (213-737-3932), holds a film festival each February that focuses on black-oriented movies. Many classics are shown. The society also holds a competition to award grants to student filmmakers, whose movies are also screened each year.

The **Natural History Museum,** 900 Exposition Boulevard, Los Angeles (213-744-3342), screens a free film every Saturday afternoon, usually a documentary coinciding with a museum exhibit. From October through May, the museum coordinates a series of free documentary screenings at area libraries, featuring everything from Mark Twain to early civilization.

New World Society, 4707 Elmwood Street, Los Angeles (213-644-4035) screens films from the Soviet Union every Friday at 8:00 P.M. as part of the group's ongoing effort to break down Cold War tensions through mutual understanding. Selections range from features to documentaries, travelogues, and cartoons.

Pasadena Filmforum, Wallenboyd Center, 301 Boyd Street, Los Angeles, holds screenings almost every Monday night. The program usually features independent filmmakers and their works, with a special jazz-film series held every October.

A couple of other screening houses generally prefer to keep a low profile, but they do exist:

Columbia Pictures Booster Club, located at Columbia Pictures in Burbank (818-954-4145), has free-admission advance screenings of upcoming films. Viewers may not be connected in any way to the film, television, or advertising industries. The club will accept local members only; no tourists need apply.

Preview House, 7655 Sunset Boulevard, Los Angeles (213-876-6600), screens television-pilot shows, commercials, and a few feature films to test public response before they're aired; audience members turn dials to indicate their likes and dislikes. Preview House prefers to invite local people only, so as not to skew marketing research. Flyers are sometimes distributed when audiences are needed. Go early, as lines can be long.

If you have an inexplicable urge for a certain type of film, be it Shakespeare, silent, or the shattering sounds of new wave, it's probably playing somewhere in Los Angeles. Dozens of art and revival houses offer specialized films, foreign and American, old and new. Some also have midnight shows of perennial favorites like the *Rocky Horror Picture Show.* Ticket prices usually are less than a first-run movie house. A brief list includes:

Art Theater, 2025 East Fourth Street, Long Beach (213-438-5435), is a neighborhood-style theater, usually offering a broad selection of foreign films and documentaries.

Beverly Cineplex, The Beverly Center, Beverly and La Cienega boulevards, West Hollywood (213-652-7760), has fourteen living-room-size cinemas under one roof, featuring art and foreign films.

Fox Venice, 620 Venice Boulevard, Venice (213-396-4215), screens revival and art movies, plus some West Coast premieres of "little" films. Also, there are special series and events, such as rock music laser shows and guest appearances by filmmakers.

New Beverly, 7165 Beverly Boulevard, Hollywood (213-938-4038), shows an eclectic range of true revivals and second-run films, plus documentaries.

Nuart, 11272 Santa Monica Boulevard, West Los Angeles (213-478-6379), presents a variety of revival, art, second-run, and documentary films, from low-budget quickies to high-budget efforts from the major studios.

Old Town Music Hall, 140 Richmond Street, El Segundo (213-322-2592), specializes in films of the twenties and thirties, with a powerful Wurlitzer organ to provide accompaniment.

Rialto, 1023 Fair Oaks Avenue, Southern Pasadena (818-799-9567), screens revival, art, foreign, and second-run films. Sometimes this theater shows double or even triple features of a filmmaker's work.

Royal, 11523 Santa Monica Boulevard, West Los Angeles (213-478-1041), is an art house that usually features first-run foreign films.

Vagabond, 2509 Wilshire Boulevard, Los Angeles (213-387-2171), specializes in revivals from Mack Sennett shorts to popular classics from the thirties and forties. This theater often shows double features of genre films—mysteries, musicals, etc.

Vista, 4473 Sunset Boulevard, Hollywood (213-660-6639), usually offers double-feature doses of revivals and foreign films, and sometimes designates a particular night of the week for genre films. Occasional special events include live comedy troupes improvising new dialogue to old movies. Zany Hollywood!

Los Angeles has many foreign-language film houses, including the **Kokusai,** 3020 Crenshaw Boulevard, Los Angeles (213-734-1148), which shows Japanese movies with English subtitles, and the **Picfair,** 5879 West Pico Boulevard, Los Angeles (213-939-5212), which shows Indian films with English subtitles. The Metropolitan theaters chain has more than thirty Los Angeles–area movie houses that feature Spanish-language films, most with English subtitles. Two of their major theaters downtown are the **United Artists Theater** at Ninth and Broadway and the **Million Dollar Theater,** built by Sid Grauman, at Third and Broadway. For information on any Metropolitan theater, call 213-624-6272.

If you like to watch your movies alone but can't afford a private screening room, the **UCLA Film & Radio Archives** and the **ATAS-UCLA Television Archives** are for you. UCLA has the second largest collection of films, television programs, and radio shows in the world, second only to the Library of Congress. It also has the only legal viewing facility for highly inflammable nitrate film stock in California, so you can see many old silent films here that otherwise couldn't be shown at all. The archives, located at 1438 Melnitz Hall on the UCLA campus (213-206-8013), are open 9:00 A.M.–12:30 P.M. and 1:30 P.M.–5:00 P.M. Tuesday through Thursday. Anyone is welcome to use the visiting and listening rooms free of charge during those times, but for no more than three hours per school quarter.

Speaking of television, **EZTV** at 8543 Santa Monica Boulevard in West Hollywood (213-657-1532) is L.A.'s first full-time video screening house, a small, comfortable room dedicated to showing new works by independent filmmakers working in video. This isn't the place to see the new "Happy Days" or "Laverne and Shirley," but it *is* the place to find everything from short works of abstract video art to feature-length, *very* low budget features. Screenings are held nightly from 6:30 P.M.–Midnight.

Finally, there's the **Cameo Theater,** 528 South Broadway, Los Angeles (213-628-1974). If fame and fortune fail you, escape to the Cameo, which offers as many as four films for as little as $1.50. Where else could you see *Under the Rainbow* and *Black Samurai* on the same bill? Open twenty hours a day, and worth a "cameo" appearance, especially if it's cold outside.

3

HOLLYWOOD OLD AND NEW

The physical history of Hollywood—studios, location sites, old mansions, and star hangouts, even a statue or two—is scattered over nearly every part of the Southland. While most of the sites are between downtown Los Angeles and the beach, there are some farther afield; Mr. Rourke's house on "Fantasy Island" is in Arcadia, while most of the planets visited by Captain Kirk and the rest of his "Star Trek" crew are out past the San Fernando Valley at a place called Vasquez Rocks.

Downtown Sites

Hollywood may have really started in downtown Los Angeles, but most of the early motion-picture sites in downtown and East Los Angeles were lost to progress long ago. The vacant lot behind Sing Loo's Chinese laundry on Olive Street, where Francis Boggs filmed *The Power of the Sultan* in 1908, is now covered with office buildings, while the site of D. W. Griffith's Biograph Studio at Girard Street (presently Twelfth Place) and Georgia Street is now the site of the Los Angeles Convention Center's parking lot. Nearby Chutes Park, which once stretched from Main Street to Grand Avenue along the south side of Washington Boulevard and provided a home base for David Horsley, the Warner brothers, and other seminal filmmakers, was subdivided after serving briefly as the home of a minor-league baseball team. The Selig Polyscope Company studio and zoo, located across from Lincoln Park at Selig Place and Mission Road in Lincoln Heights, closed for good in 1938, after serving as the first Los Angeles home of such disparate film personalities as Tom Mix and Louis B. Mayer, with the last remnant of its past, the once-famous Elephant Gates, being torn down in the early sixties.

Even the wheels of progress miss a few from time to time, though, and there are a few early—and late—movie-related sites still standing in the downtown and east-side areas.

One of the most familiar television locations of all is in the City of Industry, a town southeast of downtown Los Angeles that's one big industrial park. This is where the **McDonald's Advertising Production Studio** is located, where all the McDonald's television commercials are made. The studio, at 17030 Green Drive, looks just like any other McDonald's, except that it has two storefronts (so two commercials can be shot at the same time), a platform in the roof (so the counter can be filmed from above), dressing rooms, makeup rooms, storerooms and offices—and no customers, except for an occasional traveler who stumbles in seeking a Big Mac and finds only lights and cameras instead. Besides appearing in almost every McDonald's commercial (the impression that different restaurants are used for different commercials is created by the use of quick-change interior decor), the studio was used in *Oh God, Book II,* when George Burns—God—bought a little girl a burger.

Out in Arcadia, you'll find the **Los Angeles State and County Arboretum** at 301 North Baldwin Avenue (818-446-8251). The arboretum's 127 acres were once the center of the vast Rancho Santa Anita and were later owned by silver baron E. J. "Lucky" Baldwin. The estate, with its jungle and lake, has served as the location for numerous movies, including Johnny Weissmuller and Buster Crabbe's Tarzan films, *Devil's Island, The Man in the Iron Mask, The Women,* and *High Noon,* as well as the Jungle Jim movie serials of the forties and fifties and television's "MacArthur" and "The Letter." Some historians place the leech scene in the Humphrey Bogart–Katharine Hepburn *The African Queen* here as well.

More recently, the Queen Anne Cottage, built beside the arboretum's lake by Baldwin in 1881, has gained fame as Mr. Rourke's house on "Fantasy Island." For the show's first season, all scenes involving the house were shot here, but the high cost dictated the building of a duplicate cottage at the TBS Ranch in Burbank after that. You can't go into the cottage, except on a guided tour given twice a year, but you can see in through the windows or stand on the porch and look for "da plane, Boss, da plane." The arboretum is open 9:00 A.M.–4:30 P.M. every day except

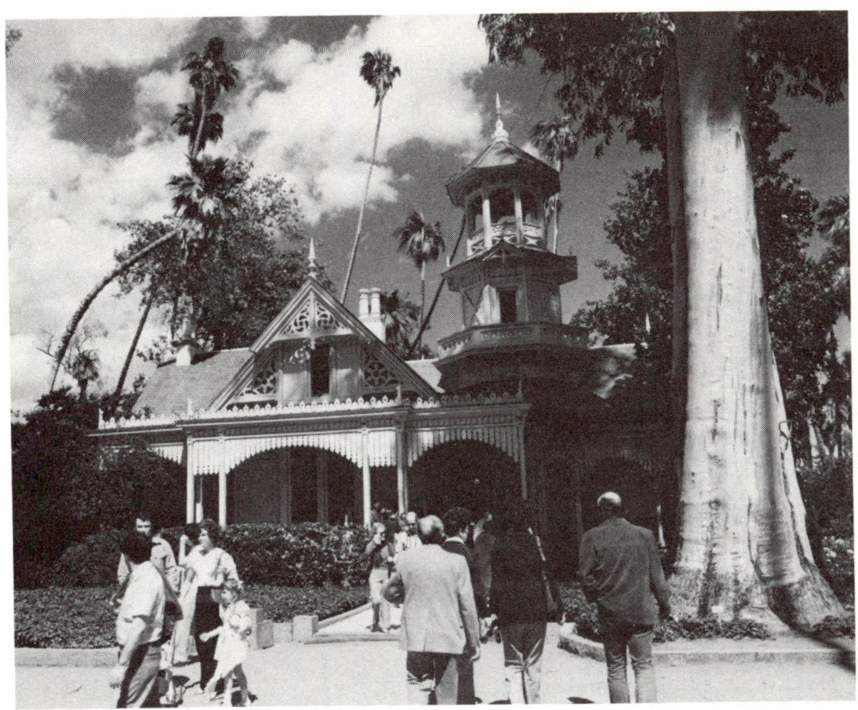

At the Los Angeles State and County Arboretum in Arcadia, anyone can stand on the porch of the "Fantasy Island" house and shout, "Da plane, Boss, da plane!"

Christmas. Admission is $1.50 for adults, 75¢ for ages five to seventeen, sixty-two and over, and students with I.D. cards, and free for children under five.

A step closer to Los Angeles is the **Huntington Library, Art Gallery and Botanical Gardens** 1t 1151 Oxford Road, San Marino (818-792-6141). Formerly the home of railroad magnate Henry E. Huntington, the Huntington Library is most famous for its collection of rare books and paintings, including Gainsborough's *Blue Boy,* Lawrence's *Pinky,* and a Gutenberg Bible. For movie buffs, though, the attraction is the Japanese Garden, which doubled for Japan in *Midway* and *The Bad News Bears Go to Japan.* The rest of the estate appeared in Peter Bogdanovich's camp classic, *At Long Last Love.* The art gallery and grounds are open free of charge 1:00 P.M.–4:30 P.M. Wednesday through Sunday, with advance reservations required on Sunday.

Just north of San Marino in Pasadena is the 1905 Fenyes Estate, now the **Pasadena Historical Museum,** at 470 West Walnut Street (818-577-

1660). Once the Finnish Consulate and now a museum of turn-of-the-century Pasadena, the eighteen-room mansion and its grounds have been used many times as a movie location, from D. W. Griffith's *The Queen's Necklace* and Peter Sellers's *Being There,* to television's "Franklin and Eleanor: The White House Years," "The Immigrants," and "Colombo." The house is open 1:00 P.M.–4:00 P.M. Tuesday, Thursday, and the last Sunday of each month for a donation of $2 for adults, $1 for students and seniors.

When filmmakers want an Oriental location, they often use Pasadena's **Pacific Asian Museum** at 46 North Los Robles Avenue (818-449-2742). The house and gardens were built in 1924 to resemble the Chinese Imperial Palace, and were used most recently in the television version of Somerset Maugham's *The Letter.* The museum is open noon–5:00 P.M. Wednesday through Sunday. Admission is $1 for adults and free for children under twelve.

One of the most famous, yet paradoxically least well-known, location sites in the Southland is the **Grand Central Air Terminal** at 1310 Air Way, Glendale. Famous, because it's where the final airport scene of *Casablanca* was filmed; little known, because it's not an airport anymore. You won't find a DC3, like the one Ingrid Bergman and Paul Henreid took—they'd have a hard time finding a place to land in the industrial park that covers the runway site today—but the airport terminal buildings that stood in the background as Rick and Ilsa said good-bye are still there.

Two oft-used ethnic locations are both in downtown Los Angeles. **Olvera Street,** a restored street of shops from Los Angeles's Spanish and Mexican period in the El Pueblo de Los Angeles State Historical Park (213-628-1274), and **Little Tokyo,** which centers on First Street between Los Angeles and Alameda streets, are used constantly by television producers looking for a little ethnic flavor. The Higashi Hongwanji Buddhist Temple at 505 East Third Street (213-626-4200) is a special favorite.

Probably the most photographed building in downtown Los Angeles is **Union Station** at 800 North Alameda Street (213-683-6987). The cavernous Spanish-and-Moorish-style station, the Los Angeles terminus of the Southern Pacific, Union Pacific, and Santa Fe railroads, opened in 1939 and was used—appropriately—in the motion picture *Union Station,* as well as *The Way We Were, St. Ives, The Driver, Under The Rainbow,* and television's "Gable and Lombard." The station, which is still in use, is open twenty-four hours.

The **Bradbury Building,** 304 South Broadway in downtown Los Angeles (213-449-3880), is a Gothic fantasy that has to be seen to be believed. On the outside, it looks like just another brownstone office building; on the inside, it's an interwoven web of cast-iron filigree stairways, open elevator cages, and marble floors, with an inner court lit by a glass roof. Besides appearing in *The White Cliffs of Dover,* a number of Boston Blackie films, and *The Cheap Dective,* the Bradbury was a prominent feature of the disturbing Los Angeles of the future in the recent Harrison Ford film, *Blade Runner.* The Bradbury is open 10:00 A.M.–6:00 P.M. Monday through Saturday, but unless you have legitimate business in one of the building's offices, the admission is $1.50 for ages twelve to sixty-five, and free for those under twelve, over sixty-five, and students with I.D. cards.

Across the street from the Bradbury is the Million Dollar Theater, the most spectacular of the eleven theaters in the **Broadway Historic Theater District** (see Guided Tours). Built in 1918 by Sid Grauman at the cost of, yes, one million dollars, the theater now shows first-run Spanish-language films from Mexico, but the baroque auditorium is worth the price of admission even if you don't speak Spanish. The Million Dollar is at 307 South Broadway (213-629-2895).

The other movie palaces in this historic district are the zigzag-moderne Roxie at 518 South Broadway; the beaux arts Arcade, formerly the Pantages, at 534; the baroque Los Angeles and the Palace at 615 and 630; the Spanish-style State at the corner of Broadway and Seventh; the beaux arts Globe at 744; the Spanish-style Tower at Broadway and Eighth; the Rialto at 812; and the Orpheum at 842. Three additional old-time movie palaces are a block north of Broadway on Hill Street: another former Pantages, now a jewelry center, at Seventh and Hill; the garishly painted Mayan at 1040 South Hill where Marilyn Monroe once appeared in a burlesque act; and the former Belasco, now the gay-oriented Metropolitan Community Church, next door at 1060 South Hill.

A few blocks east of the theater district is **Dearden's Department Store** at 700 South Main Street; at least one film historian has identified this building as the one on whose roof Selig's Francis Boggs filmed *The Heart of a Race Track Tout* in 1908, the first full-length dramatic film shot entirely in Los Angeles.

When early filmmakers and stars visited Los Angeles, they usually stayed at the **Alexandria Hotel** at 501 South Spring Street (213-626-7484), built in 1906. The glass-topped Palm Court, restored about ten years ago and now used as a banquet room, is worth a quick look. Many of the rooms and suites are named after silent-film stars, and there are photos on the walls showing the stars when they stayed there.

There are few old-fashioned ballrooms left in Los Angeles, which is why **Myron's Ballroom** at 1024 South Grand Avenue (213-748-3054) is

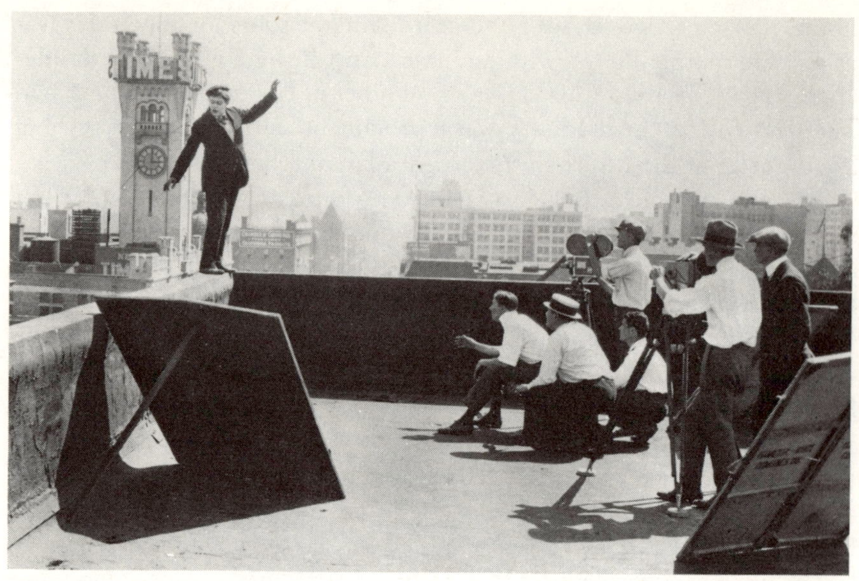

The roof of a downtown building was used as a studio in 1908 for the first, but not the last time; here, Lloyd Hamilton stars in Look Out Below *in 1922, with the* Los Angeles Times *tower in the background.*

used in movies so often. Built in 1910, Myron's numbered stars like Valentino, Mae West, and Mary Pickford among its regulars. Myron's mirrored globes and neon-lit ceiling have appeared in more than twenty films, including *Farewell, My Lovely* and *New York, New York.* Big-band dancing is still a regular feature on Fridays and Saturdays, and special events held at Myron's throughout the year include the Craft and Folk Art Museum's annual Masked Ball at Halloween.

Downtown is the home of the **Variety Arts Center** at 940 South Figueroa Street (213-623-9100), a private club devoted to vaudeville but with many film-related items, including a collection of W. C. Fields memorabilia and the Harmonia Gardens bar from *Hello, Dolly.* Some of the club's facilities are open to the public (see Attractions).

Sylvester Stallone's Rocky may have trained and fought in Philadelphia in *Rocky I, II,* and *III,* but the championship fights for all three films were actually shot in the **Olympic Auditorium,** 1801 South Grand Avenue (213-749-5171), in downtown Los Angeles.

It's now considered almost a part of downtown, but at one time the community of Edendale had a character all its own, much of it created by the many film companies who settled there, mainly along Allesandro

Avenue, now Glendale Avenue. Essanay, Selig, William Fox, and the New York Motion Picture Co. were among those who were based in the area. Today, all the studios from that period are gone, except one. In 1909 the New York Motion Picture Co., which released films under the Bison Films logo, built a studio at 1712 North Glendale Avenue. In 1912 the company turned the studio over to a new associate, Mack Sennett, who made the **Keystone Studios** famous. Besides turning out comedies featuring the famous Keystone Cops, many of whose chase scenes were filmed at the nearby intersection of Sunset and Glendale, Sennett made stars out of Roscoe "Fatty" Arbuckle, Mabel Normand, Chester Conklin, Mary Miles Minter, and an English music hall comedian named Charlie Chaplin, who started work at Keystone for $100 a week and quit a year later when Essanay offered him $1,250 a week and a $10,000 bonus. Among the productions turned out at the studio, which later covered three corners of the intersection of Glendale and Elfice, was Hollywood's first feature-length comedy, *Tillie's Punctured Romance,* a 1914 film with Chaplin and Marie Dressler that introduced the Keystone Cops.

Today, one of Keystone's original stages is the Center Theater Group's scene shop, where sets are built for stage productions at the Ahmanson Theater and the Mark Taper Forum. Call 213-662-1131 or 662-2444 if you'd like to make an appointment to look around the stage where the Keystone Cops and Charlie Chaplin used to romp.

Not far away is another equally comedic location. Remember the famous scene in Laurel and Hardy's 1932 film *The Music Box* where the boys carry a piano up a ludicrously high flight of steps only to have it roll all the way back down? The **Laurel and Hardy Steps** are still there, at 929 Vendome Boulevard, just south of Sunset in Silverlake.

A short distance from Edendale is the **S&A Studios** at 201 North Occidental Boulevard (213-384-3331) or, to give it all its previous names, the Bosworth, Morosco, Pallas, Paramount, Artcraft, Select, Realart, Famous Players-Lasky, Prizmatic, Cinecolor, Sutherland, Neufeld, Noonan-Taylor, Aldrich, VCI, and Sullivan and Associates studios. The small studio was built in 1913 by painter-actor-producer Hobart Bosworth, who had starred in Selig's *The Power of the Sultan.* Bosworth used it to make a series of films based on Jack London stories. Famous Players-Lasky (later Paramount) rented space in 1914 for their production of *Oliver Twist,* and in 1917, Cecil B. DeMille moved in under the banner of Paramount's Artcraft subsidiary and filmed Mary Pickford in *Romance of the Redwoods.* Paramount continued to control the studio for the next twenty-three years under various names, including the famous Players-Lasky Wilshire Branch, turning out films with stars such as Pola Negri, Adolphe Menjou, and a young and inexperienced Gary Cooper. For a number of years after that it was used mainly for shorts and industrial films, but in 1963 the lot went back into the feature business with the Noonan-Taylor musical *Promises, Promises,* starring Jayne Mansfield, followed a year later by *Three Nuts in Search of a Bolt* with Mamie Van Doren.

Director Robert Aldrich, trying to get away from studio interference once and for all, bought the studio in 1967 and turned out a number of films, including *The Legend of Lylah Clare* with Kim Novak and Peter Finch and *Whatever Happened to Baby Jane?* with those queens of camp, Bette Davis and Joan Crawford. Aldrich bailed out in 1973, and the studio became a rental lot once again. Recent productions have included the syndicated "Romance Theater" and portions of the features *I'm Dancing as Fast as I Can* with Jill Clayburgh, *The Hasty Heart* with Gregory Harrison, and *Twilight Zone.* The larger of the two stages on the half-square-block lot was built by Aldrich in 1968, but the smaller Stage 1 dates from the teens, although it was originally open to the sky. The office building on Occidental is basically intact, but the awnings and the fake-brick facade are obviously later additions.

In the twenties, the hotel where most stars stayed was the **Ambassador Hotel** in the Wilshire District (see Hotels). Across the street is what's left of the original **Brown Derby,** built by Hubert K. Sanborn, later Gloria Swanson's husband, in 1926. This was the Brown Derby that got the most photographs, since it's the one actually shaped like a hat, but the one in Hollywood got all the stars. The original is now demolished except for the hat, which will be moved to a new site as soon as preservation groups can find one.

Hollywood

At one time, Hollywood truly was the center of the motion-picture industry. Nearly all of the major studios were here—Fox, Warner Bros., and Columbia on Sunset, United Artists on Santa Monica, and Paramount on Melrose, with RKO on Gower Steet next door. Most of the stars were here, too, either living in the Hollywood Hills or in apartments like the Chateau Elysée and the Garden Court, or just hanging out at the myriad of restaurants and nightclubs that made Hollywood the place where everyone, not just the stars, wanted to be.

The old town's a little faded these days. All the major studios except Paramount have moved out, and a number of studios have either been converted to television use or, like the Fox lot at Sunset and Western and the Triangle lot at Sunset and Virgil (where the colossal sets for D. W. Griffith's *Intolerance* once stood), have been torn down to make room for supermarkets and parking lots. With very few exceptions, the bistros and boîtes have seen far better days; now if you meet a girl who looks like Rita Hayworth in a bar on Hollywood Boulevard, the first thing you should do is make sure she's really a girl.

Still, there's plenty to see and do in this part of movieland, and it's starting to look like Hollywood's worst is over. The town finally has a motion picture museum, a number of old Hollywood landmarks are being restored, and new restaurants and nightclubs are starting to make the area the flashy place it was in days of old. Even Warner Bros. has moved back into town, buying the old United Artists lot. This may not be Hollywood's golden age, but at least it's no longer just tarnished brass.

East Hollywood has four old movie studios still standing, two of which started as silent-film studios and are now state-of-the-art television studios.

The larger and currently more important of the two is the **ABC Television Center** at 4151 Prospect Avenue (213-557-7777). The studio was originally built by Vitagraph, one of the original film trust companies, in 1913, and was used for Westerns, comedies, and Ramon Navarro's early films. In 1925, Vitagraph was purchased by Warner Bros., and since Warner Bros. already had a brand new studio of its own in central Hollywood, the Vitagraph Studio became the Warner Bros. East Hollywood Annex. Among the early Warner Bros. films shot there were John Barrymore's *The Sea Beast* and the epic *Noah's Ark,* which killed a number of people with its re-creation of the flood. When Warner Bros. merged with First National in 1929, they kept the lot, using it mostly for exteriors for such films as Errol Flynn's *Captain Blood.* In 1949, Warner Bros. sold the studio to ABC, which converted it into what was then the world's largest television studio, and it's still ABC's main studio on the coast. The studio has been completely modernized and is rather dull to look at, but you can see the original studio gate at the corner of Prospect and Talmadge.

The smaller of East Hollywood's two television studios, **KCET Television** at 4401 Sunset Boulevard (213-666-6500), is the one with the more varied history. In 1912, a unit of the Lubin Manufacturing Co. of Philadelphia, a member of the film trust, built a studio of sorts at 1425 Fleming (now Hoover) Street, at the east end of the present KCET lot. Among the first films produced at the studio were two shorts of Los Angeles tourist attractions, *The Alligator Farm* and *The Ostrich and Pigeon Farms.* The pigeon farm must have been something to see, as this was the second film in nine years it had starred in. Of *The Alligator Farm,* the industry journal *Moving Picture World* reported, "a clearly photographed and instructive picture . . . the little curly-headed player of the Lubin company did not like to stand too close to the alligator pool." Lubin's studio closed down after a few months and eventually moved to a long-since-demolished studio in Highland Park, while "Bronco Billy" Anderson's Essanay Film Company moved into the Fleming Street studio and turned out twenty one-reel Westerns in six months.

Essanay moved out in April 1913, and in October, the Kalem Company, which already had studios in Glendale and Santa Monica, moved in. Kalem's first films at the studio were one-reel comedies like one with the tell-all title of *Fleeing From the Fleas;* the popular Ham and Bud comedies were born here in 1914, but in 1917 Kalem abandoned the Fleming Street studio for Glendale, going out of business completely before the end of the year.

The lot then became a rental facility, with one of the lessees being producer Jesse D. Hampton, who signed silent stars William Desmond and H. B. Warner to a series of films. By 1919 the lot was known as the Hampton Studio, but Hampton had already outgrown it; the following year, he moved to a new studio he had built on Santa Monica Boulevard, now the Warner Hollywood Studio.

The Fleming Street studio almost hit the big time next when one of the most popular stars of the day, Charles Ray, turned down an offer of $250,000 a year to keep working for Paramount and bought the studio. Ray built a new glass-enclosed stage, which, drastically remodeled, is still in use today as Stage 1. In 1922, Ray built the ornate Spanish-style administration building that now fronts on Sunset Drive. In preparation for his masterpiece, *The Courtship of Myles Standish,* he also built a full-scale replica of the *Mayflower,* purchased another studio at Beverly Boulevard and Virgil Street (since demolished), and announced plans for a new $3 million studio at that location.

The Courtship of Myles Standish flopped, the studio was taken over by creditors, and another Fleming Street producer had bit the dust.

After five years of almost no activity, the studio reopened as a rental studio in 1927, with most of the renters making B movies at best. In 1932, the lot was purchased by recording engineer Ralph M. Like, who remodeled Stage 1 for sound and built Stage 2. Among the stars and bit players in the many films turned out by the many shoestring companies renting space at the studio were Rita Hayworth (then called Rita Cansino) and Walter Brennan. The one and only A feature made here was John Ford's *Hurricane* in 1937.

In 1940, Monogram Pictures (the outfit to which Jean Luc-Godard dedicated *Breathless*) moved onto the lot, buying it in 1942. If it was B pictures the world wanted, then B pictures the world would get. The East Side Kids and The Bowery Boys, Charlie Chan, Jiggs and Maggie, The Shadow, and Bomba the Jungle Boy all called Monogram Pictures home. And Monogram prospered, adding offices, scene docks, a third sound stage, and a metropolitan-street set and giving a start in the business to people like Gale Storm and Ava Gardner, Roddy McDowell, Carl Foreman, and cult directors Nicholas Ray and Budd Boetticher.

Eventually, the powers-that-were at Monogram decided to spruce up their act by forming a subsidiary to produce class A product. Allied Artists (AA) was introduced in 1946, with the Monogram name dropped for good in 1952. In 1954 the company signed multipicture contracts with superstar directors William Wyler, Billy Wilder, and John Huston, but Huston's first announced project, Rudyard Kipling's *The Man Who Would Be King* with Humphrey Bogart, never made it into production. Billy Wilder's only film for Allied Artists, *Love In the Afternoon* with Audrey Hepburn and Gary Cooper, was filmed entirely in France and was only a modest success, and even Wyler's one AA film, *Friendly Persuasion,* again with Cooper, didn't return as much as it should have due to Allied Artists' poor distribution and publicity systems. The company got out of the big-budget business, turning to genre pictures like *Shack Out on 101* with a young Lee Marvin as the fry cook and enemy agent, Slob, and *Cry Baby Killer* with Jack Nicholson making his debut in the title role. Allied Artists' pledge to make "prestige pictures" was buried deep by the company's release of such sci-fi epics as *Attack of the Fifty Foot Woman* and the admittedly classier original version of *Invasion of the Body Snatchers.*

By the sixties, Allied Artists' output was downright weird. The success of *Al Capone,* starring Rod Steiger, set off a wave of crime films, and director Sam Fuller made part of his present-day reputation by making *Shock Corridor* and *The Naked Kiss,* but what can anyone say about the Vincent Price starrer, *Confessions of an Opium Eater,* or *Sex Kittens Go to College* with Mamie Van Doren?

In 1964, having all but given up production for distribution, Allied Artists moved to New York and abandoned its Hollywood studio, which became a rental lot once more. Sold in 1967 to Colorvision, it remained open for rentals, providing the stage for one of the fastest television flops of all time, ABC's 1969 "Laugh-In" spin-off, "Turn-On."

That year, perhaps as penance for its part in "Turn-On" and following virtually every other company to reside at the Fleming Street studio, Colorvision went bankrupt. Community Television of Southern California, the parent company of KCET, bought the studio in 1970 and so far has only come *close* to going bankrupt.

Today, the studio, minus the old Western and Metropolitan streets but plus a new brick-and-glass administration building, is the home of Public Broadcasting System productions ranging in scope from Walter Cronkite's "Why In the World" to Carl Sagan's "Cosmos." The old brick facade of the Charles Ray Studios is on the Sunset Drive side of the lot, and a two-hour walking tour of the studio shows off both its history and its current use (see Studio Tours).

You need a lot of imagination to picture the other two studios in East Hollywood as they were in their heyday. A large white-and-green triangle-shaped building at 1215 Bates Street, about a block from KCET, was once the **Mabel Normand–William S. Hart Studio.** Built in 1913 as the New York Motion Picture Co. Silverlake Annex, the studio originally consisted of nothing but an open platform raised above the ground. In 1916 it was turned over to Mabel Normand Productions, a company subsidiary, and when Mabel moved out in 1917, William S. Hart's company, also a subsidiary, moved in. By that time, the open platform was surrounded by wood siding topped with sliding canvas panels supported by telephone poles. When Hart moved out, the building was used for storage, and by 1926 it was derelict. Since the thirties it's been used as a theatrical scene shop. The wood siding was stuccoed over in 1938; the columns topped with laughing faces on the Effie Street side of the building contain the telephone poles that supported the original canvas diffusing system. Directly across Bates Street from the building office are two wood bungalows that were built as Normand and Hart's dressing rooms.

When Walt Disney got his first contract to supply animated Alice Comedies for a New York distributor in 1923, he and his brother Roy rented space in the rear of an office building at 4651 Kingswell Street near the Vitagraph lot. A few months later, they leased the store next door at 4649 Kingswell and put a sign in the window reading **Disney Bros. Studio.** The building is still standing, but Disney moved out in 1926 and both storefronts have long since been remodeled beyond all recognition.

Four studios in Hollywood's southeast sector have years of history behind them, and one, **Paramount Studios** at 5555 Melrose Avenue (213-468-5000), is the last major studio headquartered in Hollywood. Built in 1918 as the Paralta Studios, a rental lot, it became the Robert Brunton Studios (still a rental lot) and was the home base for William Desmond Productions, Zane Grey Pictures Corp., Mary Pickford Film Co., Mae Marsh Pictures, Raoul Walsh Productions, and dozens of other smaller companies. In 1921, it became United Studios with many of the same tenants, and in 1926 it became the new home of Paramount-Famous-Lasky Corp., later shortened to Paramount Pictures.

Paramount was essentially the creation of three men: William W. Hodkinson, an Ogden, Utah exhibitor who created Paramount Pictures Corp., the first "modern" film financing and distribution company in 1909; Jesse L. Lasky, whose Jesse L. Lasky Feature Play Co. made *The Squaw Man;* and Adolph Zukor, who created the Famous Players Film Co., and with it the first real stars. Soon after joining Hodkinson's organization along with a number of other producers, Lasky and Zukor merged, creating Famous Players-Lasky. In a quick series of moves, Zukor convinced Lasky to join him in buying out Lasky's partner and brother-in-law, Sam Goldwyn, and convinced Hodkinson's partners to kick Hodkinson out of the company he had founded and named. Famous Players-Lasky soon absorbed both the other producers under the Paramount umbrella and Paramount, as well. For the next fifteen years, Zukor (Lasky was still around, but really impotent) came close to establishing a monopoly on American film production and exhibition thanks to tactics that established new nadirs of ruthlessness. Independent theaters that refused to join the Paramount chain, for example, soon found a brand new Paramount-owned cinema rising next door. In fact, Zukor's tactics led in part to the First National-Warner Bros. and Metro-Goldwyn-Mayer mergers and the creation of United Artists.

Paramount films over the years have included the Hope and Crosby road pictures, as well as Best Picture Oscar winners *The Lost Weekend* and *The Godfather.* As famous as Paramount's films are, though, the original studio gate on Marathon at Bronson is nearly as famous; the Spanish-style wrought-iron gate is the prototype of all studio gates, caricatured in dozens of films and cartoons, and is the gate Erich von Stroheim drove Gloria Swanson through when she returned to her old studio for her comeback in *Sunset Boulevard.* Paramount's new main gate on Melrose is modeled after the Marathon gate, and it may be more impressive, but it doesn't have the history of the old one.

Around the corner at 780 North Gower Street is the side entrance to the Paramount lot; once this was the main entrance to **RKO Radio Pictures.** Neither Robertson-Cole, the British company that built the studio in 1918, nor its successor, the Film Booking Office of America, set the movie world on fire with their comedies and low-budget serials, and in 1926 Radio Corporation of America (RCA) bought the FBO Studios in a deal arranged by Joseph P. Kennedy, merging it with Pathé Pictures and the Keith-Albee-Orpheum theater circuit to create Radio Keith Orpheum, better known as RKO. RKO used its Hollywood lot mainly for interiors for such films as Fred Astaire's *Top Hat* and *Flying Down to Rio,* the Marx Brothers' *Room Service,* and *Topper* with Cary Grant, while their Culver City studio was used mostly for back-lot exteriors. Howard Hughes bought RKO in 1948 and sold it in 1955 after making a handful of films, the last of which was *Jet Pilot* with John Wayne. The following year, Lucille Ball and Desi Arnaz bought both RKO lots for their Desilu company, with the Hollywood lot being used for television shows from "I Love Lucy" and "The Real McCoys" to "Hogan's Heroes" and "Ben Casey." Paramount bought the lot from Desilu and merged it into their own in 1967. The globe on the top of the stage at the corner of Sunset and Gower is a reminder of the RKO period; it once served as the base for a stylized radio tower, RKO's logo.

Raleigh Studios, roughly across the street from Paramount at 650 North Bronson Avenue (213-466-3111), began as the Clune Studio and Laboratories in 1915, built by Los Angeles theater owner William H. Clune. Clune turned out a number of epics, beginning with the original version of *Ramona,* before turning the studio over to a variety of renters, including Douglas Fairbanks, who made *The Mark of Zorro* and *The Three Musketeers* here. It has remained a rental studio under a variety of names; as the Tec-Art Studios, it was used by Walt Disney for overflow animation work, and as the California Studios and Producers Studio it was used for Hop-along Cassidy serials and "Superman" television episodes, as well as most of the Roger Corman-American International Pictures' Edgar Allen Poe films and Stanley Kramer's *In the Heat of the Night* and *The Pride and the Passion.* It's presently used mostly for commercials, although Faye Dunaway's "Evita Peron" television movie was filmed there as well. The original entrance is around the corner from Bronson at 5620 Clinton Street.

North of Paramount at 5823 Santa Monica Boulevard, across the street from the Hollywood Memorial Park, is a now-closed Spanish-style building that was one of the funniest places in town. Established in 1916 as the **Pacific Studio and Laboratory** and taken over as the National Film Co.

Annex in 1919, the place started out normally enough, with part of the original Elmo Lincoln *Tarzan* being shot here. Then in 1921 it became the Berwilla Studio, a rental lot used almost exclusively by comedy producers. As the Darmour Studio from 1930 until the 1940s, it remained a comedy lot, with Louise Fazenda comedies among its productions. Darmour released through Columbia, and when Larry Darmour died in the forties, Columbia bought the studio for the Columbia Annex, where The Whistler and The Lone Wolf serials were filmed, along with Blondie and The Three Stooges. Columbia sold it off to Family Films in 1950, which turned it back into a rental lot. It was sold again a few years ago to a real-estate developer who started restoring it; then work abruptly stopped. Today it's empty, except for the guard dog in the parking lot.

The studios continue in south-central Hollywood, between Santa Monica Boulevard and Melrose Avenue west of Vine Street. **Television Center Studios** currently owns two historic locations, the former Technicolor lab at 6311 Romaine Street (213-462-5111) and the former Cinema General-Motion Picture Studios at 1035 Cahuenga Boulevard. The studio was built as the Equity Studios rental lot in 1946 and became the Motion Picture Center Studios rental lot in 1947, with Stanley Kramer making it his base in 1949 for *Home of the Brave* with Kirk Douglas and parts of *High Noon* in 1952. Desilu started leasing space on the lot in 1953, using it for some early episodes of "I Love Lucy," "Our Miss Brooks," and the Danny Thomas and Jack Benny shows. Desilu bought control of the studio in 1955 and the remaining 49 percent in 1957, shooting "The Honeymooners," "The Dick Van Dyke Show," "Make Room for Daddy," and "The Real McCoys" on the lot and renaming it the Desilu Cahuenga Studios. When Desilu sold out to Paramount in 1967, the studio went along as part of the deal, but was resold a short time later to Cinema General Studios, which in turn sold it to Television Center Studios, which continues to operate the lot as a rental studio, with "General Hospital" and "Soap" among the recent tenants. Presently the lot includes the Hollywood Playhouse Theater on Stage 9, where live stage shows are taped for cable television.

The Television Center Studios parking lot on Romaine, by the way, was once the Metro Pictures lot before the merger with Goldwyn that created MGM.

Presently, the most famous—or at least most notorious—studio in this part of Hollywood is Francis Ford Coppola's **Zoetrope Studios** at 1040 North Las Palmas (213-463-7191). The lot was designed and built in 1919 as Jasper Hollywood Studios, a "unit" studio that contained a number of stage and office combinations that were rented out to independent filmmakers. Harold Lloyd and King Vidor were two of the early tenants. The studio fronted along Santa Monica Boulevard until 1925, when part of the land was subdivided and the entrance for renters was moved around the corner; the lot, then called the Hollywood Metropolitan Studios, was owned at that time by the Christie Brothers' Metropolitan Productions, who had their own entrance, now closed, on the south side of the lot at 6625 Romaine. In spite of their success with Bert Lahr and the Ritz Brothers, the Christies went bankrupt in 1932, and the receiver, General Services, Inc., put their own name on the lot.

In the late thirties, the lot was used extensively by United Artists. In the fifties, the Nasser brothers, (*not* the Nassour brothers, who had built the lot on Sunset that's now Metromedia Square) took over the lot, renamed it Hollywood General Studios, and held onto it until Coppola came along with his dream of reestablishing the studio system single-handedly, complete with stock players on long-term contracts. That dream, though, proved to be beyond even Coppola's generous reach. For *One From the Heart,* Coppola re-created Las Vegas inside a sound stage at a cost of $4 million; when the film, which cost a total of $27 million, crashed, returning just $1 million at the box office, it took Zoetrope Studios with it. A foreclosure auction was scheduled for Valentine's Day 1982—one from the heart, indeed—but Coppola pulled one from the hat and held off the sale for a short time by announcing he had found two possible mystery buyers. As the saying goes, stay tuned. You can still look through the gate, though, and see the Las Vegas airport facade built in front of one of the sound stages for "Heart," as well as street signs with names like Federico Fellini Drive.

A little west of Zoetrope at 7200 Romaine Street is a small art deco building that wouldn't draw a second glance from most people; Howard Hughes liked it that way when he ran his **Caddo Productions** from this spot.

Warner Hollywood Studios at 1041 North Formosa Avenue (213-850-2500) was built by producer Jesse D. Hampton in 1919 and sold to Mary Pickford and Douglas Fairbanks in 1922 for their fledgling United Artists (UA) company. United Artists was just what its name implied, a company owned by the artists themselves as a way to control their films. Founded in 1919 by Charlie Chaplin, D. W. Griffith, and William S. Hart, along with Pickford and Fairbanks, United Artists, had by the end of the twenties, taken in Gloria Swanson, Sam Goldwyn, Howard Hughes, and Buster Keaton as well. Fairbanks made the first major films on the lot, *Robin Hood* and *The Thief of Bagdad,* but Goldwyn was the most prolific producer on the lot, turning out hits like *Stella Dallas* and *Wuthering Heights.* He also proved to be as conniving as Adolph Zukor. Little by little he bought more and more of the lot, so that by the time he left the company in 1941, UA was just a tenant on the Samuel Goldwyn Studios lot. Goldwyn, whose post-UA films included *The Best Years of Our Lives, Guys and Dolls,* and *Porgy and Bess,* stopped making pictures in 1959 and the studio became a rental lot, used mainly for television shows such as "Barnaby Jones." Goldwyn died in 1974, and in 1981 his heirs sold the studio to Warner Bros., with the proceeds going to the Motion Picture-Television Relief Fund. The lot is still used mostly for television production; UA has relocated on the MGM lot since its purchase by MGM owner Kirk Kerkorian.

As full as Hollywood is of studios, there's more to the heart of movieland than film factories alone. The first movie mansions anywhere in Hollywood were in east Hollywood, in the expensive Laughlin Park subdivision just south of Griffith Park. **Cecil B. DeMille** started the movie colony's move toward grandeur when he moved out of his cabin in Cahuenga Pass and paid $28,000 for a Laughlin Park mansion on a street that was quickly renamed DeMille Drive. Two years later, the just-married **Charlie Chaplin** moved into the house next door. DeMille later bought Chaplin's house to use as offices and guest rooms, and the two structures, at 2000 and 2010 DeMille Drive, are now connected. DeMille died in 1959 and the house is now the offices of the DeMille Foundation, with DeMille's office still maintained just as it was when he was alive.

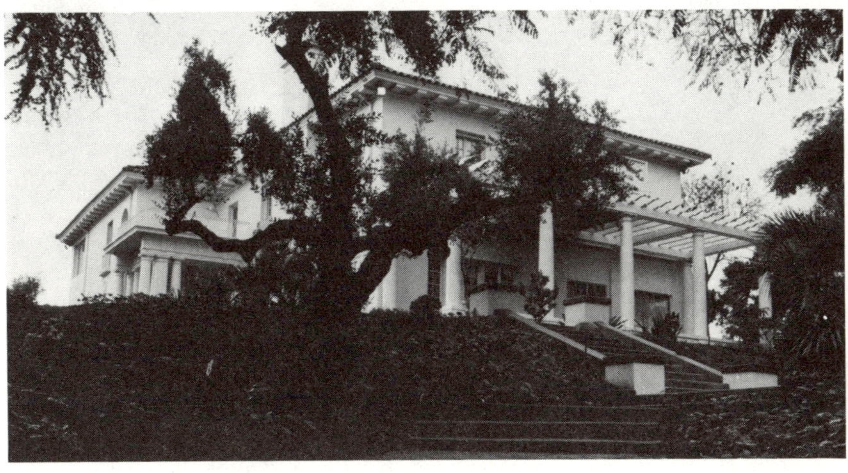

The first of the real Hollywood mansions: Cecil B. DeMille's mansion in Laughlin Park.

DeMille and Chaplin weren't alone in choosing Laughlin Park as a place to live. For a time in the forties, **W. C. Fields** lived just across the street at 2015 DeMille Drive (now owned by Lily Tomlin), and at the very top of the hill at the end of Linwood Drive is **Carole Lombard's** art deco palace. All the streets in Laughlin Park are privately owned, so drive discreetly; no parking is allowed on the streets.

Across Los Feliz Boulevard from Laughlin Park, just to the west of the Fern Dell section of Griffith Park, is proof that not all movie stars were concerned just with opulence. Some were into art as well, as evidenced by silent-film star **Ramon Navarro's** Mayan fantasy of a house at 5609 Valley Oak Drive. Designed in the twenties by Lloyd Wright, the son of Frank Lloyd Wright, the small house has outrageous jade-green metal trim and startles the hell out of lost drivers who don't know it's there.

Just below Fern Dell is the present West Coast home of the **American Film Institute (AFI)** at 2021 North Western Avenue (213-856-7600). AFI took over the campus of Immaculate Heart College in 1981, turning it into offices, a fine film library, screening rooms, and classrooms, sound stages, and editing rooms for students of AFI's Center for Advanced Film Studies, an elite post-graduate film school for directors, writers, produc-

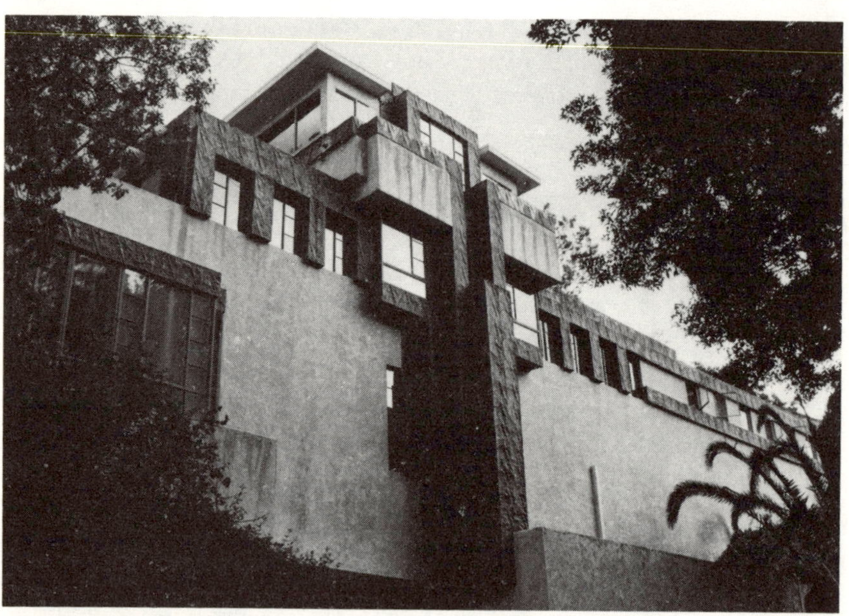

Most of the Hollywood stars chose Spanish-style mansions in Beverly Hills, but some, like Ramon Novarro, took a more radical approach.

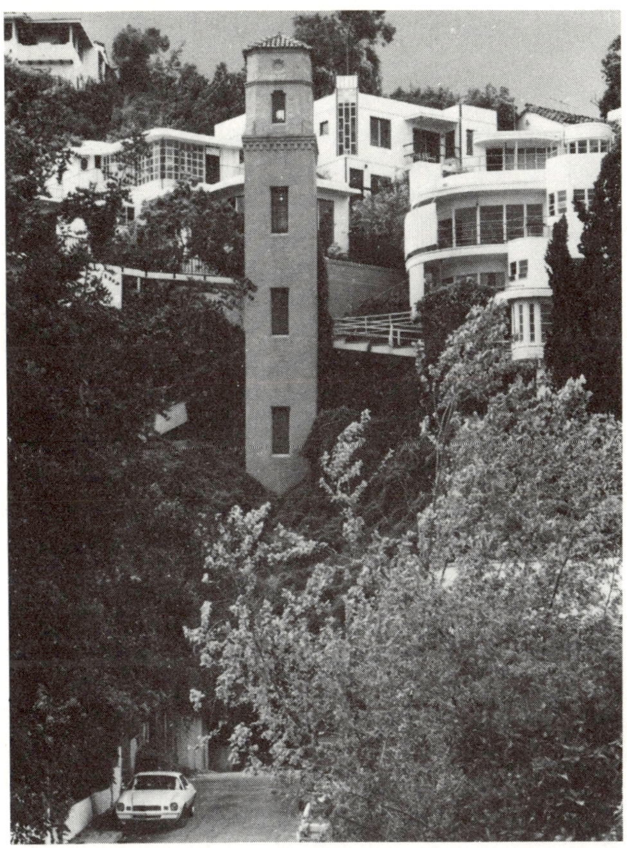

The Hollywood Hills hides one surprise after another, here the appropriately named High Tower. In Robert Altman's version of Raymond Chandler's The Long Goodbye, *Elliot Gould's Philip Marlowe lived in the house with the curved windows to the tower's right.*

tower was the most interesting thing about *The Long Goodbye,* which starred Elliott Gould as Philip Marlowe to the horror of Bogart fans everywhere. In that film, Gould lived in the art deco house to the right of the tower (as seen from below), was entranced by the girl who lived in the house immediately on the other side of the tower, and had a lot of trouble with his cat.

The Japanese palace visible at the top of the hill behind the Chinese Theater is the **Yamashiro Skyroom,** built in 1914 as a residence by two Oriental-product importers, the Bernheimer brothers. Now a bar and restaurant with so-so Chinese and Continental food and the most spectacular view of any restaurant in Los Angeles, the building and its Japanese gardens doubled for the real thing in *Sayonara* and *The Teahouse of the August Moon.*

Below Yamashiro's at 7001 Franklin Place (213-851-3313) is **The Magic Castle,** also once a private house, now an extravagantly Victorian private club for magicians. If you can get in, you'll meet such regulars as Irma the invisible piano player, who'll drink your drink if you're not careful, as well as the very visible magicians who perform nightly in the different rooms. The club is used as a location for television shooting nearly every week; whenever you see a magic act on a television series, the odds are it was shot here.

Besides being the main street of Hollywood, **Hollywood Boulevard** is also a former film location in its own right. In Paul Mazursky's *Alex in Wonderland,* Donald Sutherland played a film director who fantasized staging a gory street battle on Hollywood at Vine; of course, it was Mazursky who actually staged the battle that Sutherland's character only dreamed about.

Along with Sunset Boulevard and Vine Street, Hollywood is also the location of the **Walk of Fame** (see Attractions), which immortalizes more than seventeen hundred radio, television, film, and theater personalities in terrazzo and brass.

Hollywood Boulevard is no longer the tony street it once was; celebrity hangouts and expensive stores have given way to slimy bars and sleazy shops. It hit bottom some time ago, however, and is gradually but surely on the upswing, and there are a number of remnants of a better time stretched along its course.

The **Florentine Gardens,** 5955 Hollywood Boulevard (213-464-0706), is one of the few major nightclubs from the forties that hasn't been torn down. Never as popular as the showrooms on the Sunset Strip, the club featured such acts as Sophie Tucker, the Mills Brothers, and the Flying Herzogs trapeze act before closing in 1954. It recently reopened and is used for punk rock and other dances from time to time.

Al Jolson was the MC at the opening ceremonies when the 2,800-seat **Pantages Theater,** at 6233 Hollywood Boulevard (213-462-3104), opened as Los Angeles's most spectacular movie palace in 1930. The Academy Award presentations were held here from 1949 to 1959, but movies moved out and live stage acts moved in in 1977. The Pantages is the largest and grandest theater in Hollywood, with a bronze-and-marble art deco interior that has to be seen to be believed. The theater is used mainly for musical acts, from *Bubbling Brown Sugar* and Bob Fosse's *Dancin'* to Peter Allen and the Rockettes.

Its first show upon opening in 1926 was *Charlot's Revue,* a British music hall show with Beatrice Lillie and Gertrude Lawrence.

Next door to the Paramount at 6840 Hollywood Boulevard is the **Masonic Temple.** The beaux arts building was erected in 1921, and D. W. Griffith's funeral was held here in 1928.

The most famous movie palace in the world, and justifiably so, is the **Chinese Theater** at 6925 Hollywood Boulevard (213-464-8111). The outrageous Chinese decor and the famous footprints and signatures of the stars in the cement forecourt make Sid Grauman's 1927 creation a Mecca for tourists. Contrary to legend, Norma Talmadge wasn't the first to be immortalized. When the court was being laid, Grauman walked across some wet cement and got both a lecture from his head mason and an idea. He called up Douglas Fairbanks, Mary Pickford, and Norma Talmadge and had all three step in the cement by the curb, but the cement was almost dry and the impressions didn't take, so he called them back to do it again in a special slab of fresh cement in the center of the court. Fairbanks and Pickford went first, followed by Talmadge a few days later, just before the theater opened with the premiere of Cecil B. DeMille's *King of Kings.*

Today, the courtyard of the Chinese Theater has the prints of more than 150 stars, including Betty Grable's leg print and the hoof prints of

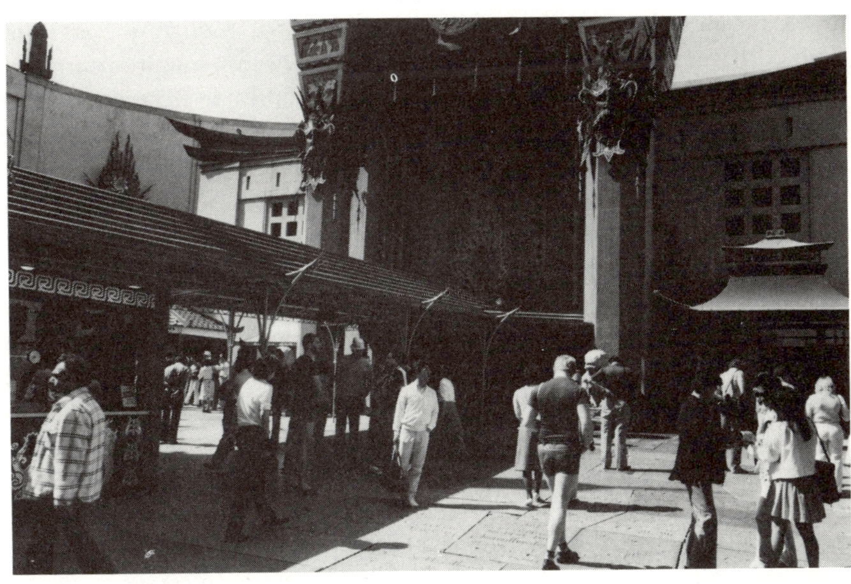

Exotic as the Chinese Theater may be by itself, it's the forecourt with its dozens of stars' footprints in cement that's made the Chinese the number-one tourist attraction in town.

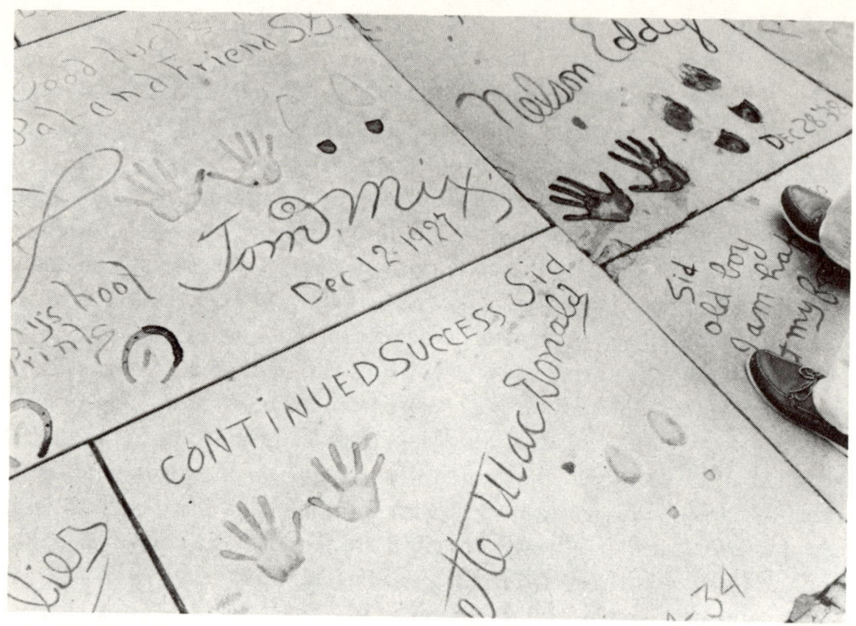

Trigger, Roy Rogers's horse. The best way to see the lavish inside of the theater is to buy a ticket to whatever's playing and experience it the way it was meant to be experienced. If you're in a hurry, though, the StarLine tours office in the courtyard sells tickets to a twenty-minute tour, conducted by a theater staff member. You'll see the theater close up, but the information given varies in accuracy depending on how knowledgeable the person giving the tour is, and whether he's bored enough to start making things up. The tour is given only in the mornings, with times varying according to the theater's schedule, and costs $2.50 for ages twelve and over and $1.50 for ages five to eleven, with no charge for children under five.

If you're not hungry enough—or rich enough—to dine at Musso and Frank's, you can still get a taste of old Hollywood by stopping in **C. C. Brown's** chocolate shop and ice cream parlor at 7007 Hollywood Boulevard (213-462-9262). Opened in 1929, C. C. Brown's is best known for being the place where the hot-fudge sundae was born. Not only did the stars get their ice cream here, but one was created here in 1936 when agent William Demarest (later Uncle Charlie on "My Three Sons") spied a girl selling chocolates and took her to Paramount for a screen test. The girl, Ellen Drew, became a minor star in the forties, but not before returning to Brown's to create the Cinderella sundae in her own honor.

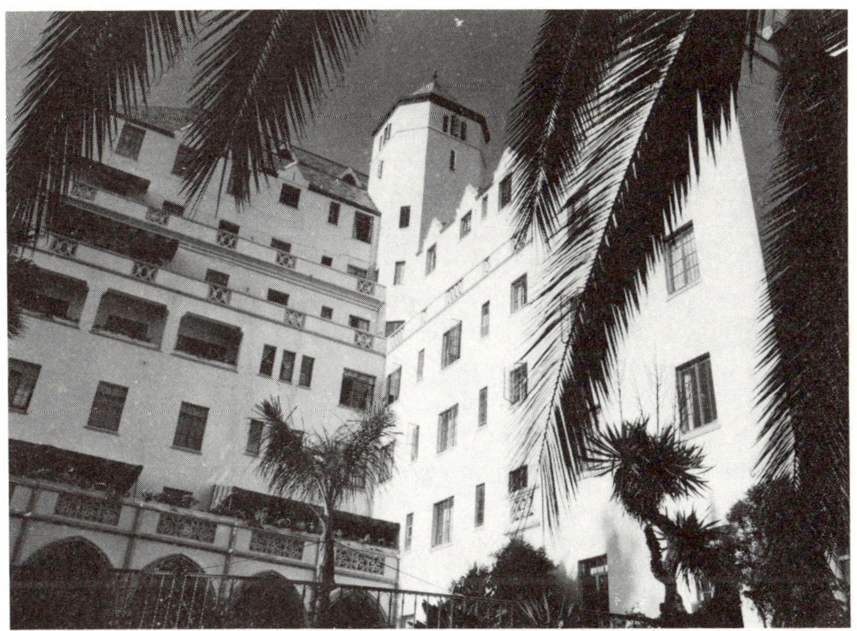

Chateau Marmont on the Sunset Strip; Karloff, Garbo, De Niro, and Belushi slept here.

has provided a refuge since 1925 for everyone from Boris Karloff, Greta Garbo, and Howard Hughes to Robert De Niro. You're welcome to look around the lobby, which looks more like the reception area of a private girl's school dorm than a fancy Hollywood hotel.

Across the street at 8218 Sunset is a large statue of Bullwinkle J. Moose and Rocky the Flying Squirrel that marks the home of **Jay Ward Productions,** once the originator of such memorable characters as Crusader Rabbit and Rags, Boris and Natasha, Mr. Peabody and his boy Sherman, George of the Jungle, and Super Chicken, but now reduced to turning out sixty-second shorts featuring an elephant named Smedley and a seafarer (and cereal) named Captain Crunch.

Most of the restaurants and nightclubs on the Strip have some kind of history behind them. The **Imperial Gardens** Japanese restaurant, 8225 Sunset (213-656-1750), was built as the Players by director Preston Sturges; **The Source,** 8301 Sunset (213-656-6388), a health-food restaurant built by one of Los Angeles's more obscure cults, is where Woody Allen broke up with Diane Keaton in *Annie Hall;* the **Comedy Store** at 8433 Sunset (213-656-6225), the nightclub that gave a start to such comics as Jimmie Walker and David Letterman, was originally Ciro's, where the acts in the forties ran more toward Martin and Lewis, Sammy Davis, Maurice Chevalier, Josephine Baker, and Peggy Lee; and **Butterfields** restaurant, a small wooden cottage with an outdoor patio hidden below the street at 8426 Sunset (213-656-3055), was reputedly once the guest house of an estate where John Barrymore lived before building Belle Vista.

A block below Sunset, next to the art deco Sunset Towers Apartments (which is being restored in fits and starts) is a small, wood-frame colonial-style house at 8341 DeLongpre Avenue that was once the home of silent Western star **William S. Hart** and now houses **Actor's Studio.** The house itself is open only to students and theater audiences, but the small lawn is open as a little-maintained, vest-pocket county park. There are a couple of picnic tables, and it's a good place to bring a bag lunch and get away from the bustle of Sunset Boulevard. You might also get a glimpse of some of the well-known actors who still return to the Actor's Studio for refresher classes and scene work.

The currently-closed restaurant at 8524 Sunset Boulevard was once **Dino's Lodge,** owned by Dean Martin, and the door to the upstairs offices of the Tiffany Theater next door was famous for years as the door to "77 Sunset Strip." Ed "Kookie" Byrnes parked his cars in Dino's parking lot.

Besides Ciro's, the two swankiest nightclubs of the thirties and forties were the **Mocambo** at 8588 Sunset and the **Trocadero** at 8610 Sunset. Don't bother looking for either of them, though; all you'll find today are parking lots. Today, the only nightclub on the Strip that regularly attracts stars is the **Roxy** at 9009 Sunset (213-276-2222), and it's more likely to host rock stars listening to other music acts than it is to host movie stars.

Next door to the Roxy is the **Rainbow Bar and Grill** at 9015 Sunset (213-278-4232). Like the Roxy, it's a rock-and-roll hangout—the parking lot at 2:00 A.M. is a real zoo—but in the forties and fifties it was the Villa Nova Restaurant, where Vincent Minnelli proposed to Judy Garland.

At the bottom of the hill at the west end of the Strip is **Chasen's** restaurant (9039 Beverly Boulevard, 213-271-2168), where W. C. Fields used to play Ping-Pong in the back room while Errol Flynn sat in the restaurant's steam room. For years this was one of the most star-studded restaurants in Los Angeles, counting Howard Hughes, George Raft, Cary Grant, and Jimmy Cagney among its regulars. It's still popular with the older generation of stars, who keep going for the sedate old-line atmosphere and Dave Chasen's chili, which isn't on the menu and which is certainly the most expensive chili ever invented.

Chili was also one of the attractions at **Barney's Beanery** at 8447 Santa Monica Boulevard (213-654-2287). While Cagney and Grant were dining in style at Chasen's, John Barrymore, Jean Harlow, and Clara Bow were getting their meals in this dive, built as a house around 1910 and opened as a roadhouse in 1920. Only the bar is in its original state, although the various enlargements did little to disrupt the colorfully seedy ambience. Barney's was famous twice in the sixties, once when artist Ed Keinholz created a life-size replica of the bar in resin for a show at the Los Angeles County Museum of Art, and once again when it was picketed by gay-liberation groups for refusing to take down the Fagots *(sic)* Stay Out sign. The sign's still there.

At the southeast edge of West Hollywood is **CBS Television City** at 7800 Beverly Boulevard (213-460-3000). Built in 1952, the building is still CBS's main West Coast studio. Previously, this was the site of Gilmore Field, home of the Hollywood Stars baseball team of the old Pacific Coast League. The team's owners included George Burns and Gracie Allen, Walt Disney, C. B. DeMille, Gary Cooper, and Bing Crosby.

Beverly Hills

Just as no place is more strongly associated with the movies than Hollywood, no place is more strongly associated with movie stars than Beverly Hills. The name conjures up instant images of wealth and luxury far beyond most people's modest dreams; "The Beverly Hillbillies" by any other name would not have smelled so rich.

From the very start, Beverly Hills, which was named by Burton Green, the president of the Rodeo Land and Water Company that developed the city, was planned as a strictly upper-class place. In 1907, three years after the first home was built in Beverly Hills, only six houses stood in the town; it took the construction of the Beverly Hills Hotel in 1912 to convince the swells that maybe the new city wasn't too far out in the sticks after all.

Entering Beverly Hills from the Sunset Strip, the city limits are immediately apparent as nightclubs and office buildings give way to tree-shaded streets and mansions with broad expanses of lawn. The newest celebrity neighborhood is Trousdale Estates, which was developed in the sixties. Many of the homes here are garish pseudo-reproductions of Greek and Roman villas, complete with columns and statuary but completely lacking in class. One of the gaudiest and worthy of special mention is **Danny**

Thomas's Villa Rosa at 1187 Hillcrest. The round chapel in the rear of the house, visible from farther down the hill, is where daughter Marlo exchanged wedding vows with talk-show host Phil Donahue. Much more sedate is the former home of the late **Elvis Presley** and Priscilla Presley at 1174 Hillcrest; the iron gate is covered with messages of undying love scratched in by fans. Down the hill at 1120 Wallace Ridge is the former home of **Warren Beatty;** after Beatty moved out, the home was used to portray Jack Warden's home in Beatty's *Shampoo.*

Today, Trousdale is covered with homes, each selling for seven figures, but at one time all of Trousdale was just the private orchard of oilman Edward C. Doheny, whose **Greystone Mansion,** at 501 North Doheny Road, is the largest and most opulent estate ever built in Beverly Hills, bar none. The 46,000-square-foot, fifty-five-room mansion, with its own bowling alley, took three years to build and was completed in 1928 at a cost of $4 million. The rest of the 410-acre estate included stables, tennis courts, a swimming pool (now filled in) with a combination dressing room and bandstand, and formal gardens galore.

Greystone has been the site of much film location shooting. The entire interior was painted green for *The Loved One;* traces of the paint are still visible on the once-magnificent woodwork. Other productions shot on

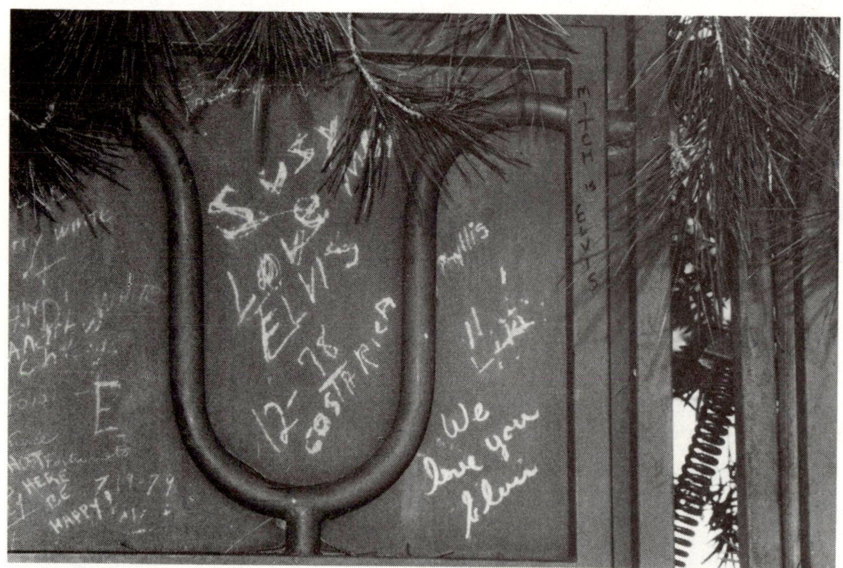

Stars may die, but fame lingers on, as these messages from fans scribbled on the gate of Elvis Presley's former home prove.

the grounds or in the mansion include *Winter Kills, Johnny Got His Gun,* and the television mini-series "The Bastard."

In 1969, five years after the mansion and the twenty-five acres around it were sold to the city of Beverly Hills, the mansion became a film location of another sort when it was leased to the American Film Institute as its West Coast headquarters and the AFI Center for Advanced Film Studies (see Attractions). The institute moved out in 1981, and today the mansion is vacant. The grounds are open daily, free of charge, 10:00 A.M.–5:00 P.M. in the summer and 10:00 A.M.–4:30 P.M. in the winter as a city park, and during the summer a number of special events are held there, including Sunset Concerts in the courtyard one Sunday a month and the Summer Solstice Dulcimer Festival in late June. For more information, call 213-550-4864.

Doheny wasn't the only oilman to settle in Beverly Hills; Max Whittier, one of the city's founders, had the same idea, and he built a thirty-eight room mansion on a 3.5-acre site at the corner of Sunset Boulevard and Alpine. In 1978, however, the Whittier mansion became known forevermore as **The Sheikh's Mansion** when it was purchased for $2.5 million by twenty-eight-year-old Saudi Arabian Sheikh Mohammad S. A. Al Fassi and his twenty-two-year-old wife, Sheikha Dena. The sheikh proceeded to shock his new neighbors by painting the white house gas-chamber green, installing a copper roof, planting plastic flowers in the flowerpots on the front wall, and painting the naked female statues on the veranda. Some of the statues received amazingly natural paint jobs, while others ended up looking like the result of a bad dream shared by Hugh Hefner and Salvador Dali. Tourists loved it; Beverly Hills was not pleased. The Sheikh left town shortly thereafter, on the run from creditors and the ex-Sheikha Dena's divorce lawyer. In 1980, the Sheikh's mansion was seriously damaged in a fire allegedly set to cover a burglary by a former employee, and in late 1982, the security guards hired to watch the place quit, after posting signs calling the Sheikh a deadbeat.

Oh, yes. The house does have a movie connection. Steve Martin's *The Jerk* was partly filmed at the mansion. The producers were looking for a place that looked like it had been built by someone who had suddenly come into great wealth but who had the worst taste in the world and who had furnished his estate with the kitschiest dreck from four continents. They didn't have to look far.

The original stars' houses in Beverly Hills were on the flats, the areas just north and south of Sunset Boulevard. Today, the flats are the low-

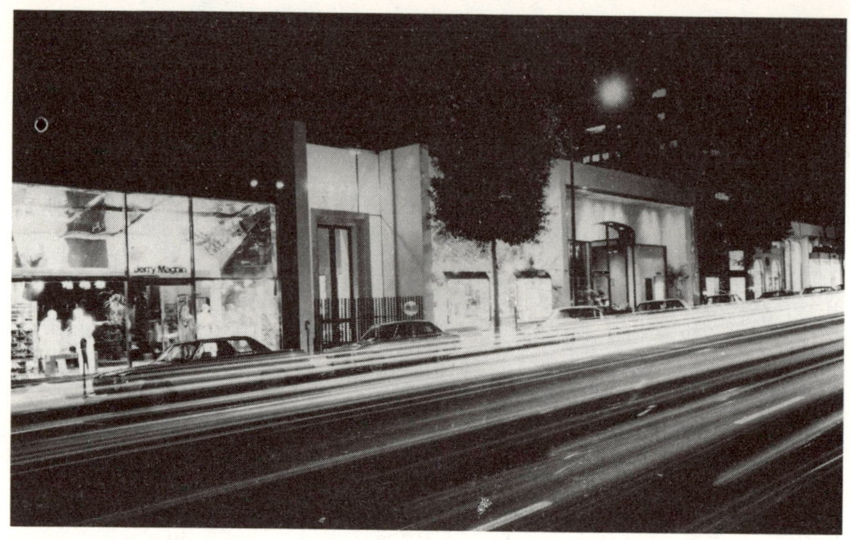

Day or night, Rodeo Drive is the glitziest shopping street in Beverly Hills, with Gucci, Hermes, and other high-buck shops catering to the famous, the infamous, and the merely rich.

rent district as far as Beverly Hills is concerned. Most of the houses in the flats aren't really mansions; they're just big houses that would sell for a fifth of their million-dollar-plus prices in any eastern or midwestern city. Few deserve special attention. At 920 North Maple is the former home of **George Burns** and **Gracie Allen.** Why single it out? Because the house in "The George Burns and Gracie Allen Show" was an exact duplicate, except for the street number.

Of the many churches in Beverly Hills, the one that many people find most familiar is **All Saints Episcopal Church** at 504 North Camden Drive, the corner of Camden Drive and Santa Monica Boulevard, (213-272-3742). It was at a wedding in this church that Dudley Moore first spied Bo Derek, the "10" of his dreams.

The flats are also the location of the number-one star-seeing shopping area, the so-called Golden Triangle, bordered by Wilshire and Santa Monica boulevards and Rexford Drive and roughly bisected by **Rodeo Drive.** Besides Gucci, Hermes, Giorgio, and the rest of the usual haute couture shops, Rodeo is where you'll find the rich and the ultrarich shopping for such necessities of life as a $10,000 gold-plated .38 caliber revolver at Bijou. At Wilshire and Rodeo is the **Beverly Hills Hotel** (see Hotels).

It's not on Rodeo Drive, but **The Bistro** at 246 North Canon Drive

(213-273-5633) is as classy as any restaurant that is. *And,* it's not only been a movie location, but a movie set as well. The gay nineties Parisian interior, all art nouveau wood and glass, was built from the set of *Irma La Douce,* and the upstairs bar was the scene of the election-night celebration scene and Julie Christie's unforgettable offer to Warren Beatty in *Shampoo.* The menu is country French, but from the elegant and expensive part of the country.

Where Rodeo Drive hits Sunset, you'll find the small **Will Rogers Memorial Park,** one of four parks in the Los Angeles area named for the humorist-trick roper-movie star who was the first honorary mayor of Beverly Hills. Across from the park is the fabled Pink Palace, the Beverly Hills Hotel.

Most of the movie palaces in Beverly Hills are in the hills north of the Beverly Hills Hotel. Probably the most famous is **Pickfair,** the home of Mary *Pick*ford and Douglas *Fair*banks, at 1143 Summit Drive. Originally built in 1911 as a six-room hunting lodge with no electricity or running water by attorney Lee Phillips, Fairbanks bought the house in 1919 and had his art director, Max Parker, turn it into a huge Tudor mansion more suitable for a Hollywood star. By the mid-twenties, "Doug and Mary" were entertaining so many royal visitors, beginning with Alfonso XIII, the last king of Spain, that they became known as the King and Queen of Hollywood. Then Pickford's career hit the skids as she outgrew her Little Mary role, she and Fairbanks were divorced, and Mary married Buddy Rogers. She became a recluse, rarely leaving her second-floor bedroom, and in 1979 she died. Today, Los Angeles Lakers owner Jerry Buss owns the house, which the press, never ones to miss a cheap joke, have redubbed "Bussfair."

Pickfair is hardly visible these days through the high shrubbery and trees, but Harold Lloyd's **Greenacres,** on the other side of Benedict Canyon at the top of Greenacres Place, is, thanks to the recent subdivision of most of its grounds. If Pickfair was the first real movie-star mansion in Beverly Hills, Greenacres, completed in 1929 at a cost of $2 million, was the largest and the grandest. The forty-room, 36,000-square-foot Spanish-style house was just the center point of the estate, which featured a nine-hole golf course, an 800-foot-long canoe pond fed by a 125-foot-high waterfall, and formal gardens resembling an ancient Roman emperor's estate. In fact, the gardens were used to portray exactly that in the orgy scenes in the Yul Brynner movie *Westworld.*

Lloyd died in 1971, and Greenacres briefly became the Harold Lloyd Museum, but complaints by neighbors shut it down after a year of operation in 1974. While most of the sixteen-acre estate has been subdivided, with new houses stretching along Benedict Canyon below the house, the house itself was purchased and restored in 1980 by art collectors Bernard and Dova Solomon. With most of the landscaping gone, the Lloyd house is very visible from the top of Greenacres Drive, the estate's former driveway, or from below on Benedict Canyon Road. The remaining fountains and terraces, as well as the size of the mansion, give some hint of the glory that was.

John Barrymore's Benedict Canyon estate, **Belle Vista,** has also been subdivided, so that today you have to go to three different addresses to see all of it. When Barrymore got tired of living at the Ambassador Hotel with his pet ape, Clementine, he bought King Vidor's small Spanish-style house at 1235 Tower Road. He then built another Spanish-style house up the hill, naming the first house Liberty Hall and the new house the Marriage House when he married Dolores Costello in 1928. By 1938, Belle Vista had sixteen buildings and fifty-five rooms on seven acres, including an aviary whose star tenant was a vulture named Maloney. The aviary has since been turned into a house whose tenants have included Katharine Hepburn, Candice Bergen, and Marlon Brando. No doubt there are some cynics who would have preferred the vulture.

The present address of the Marriage House is 9876 Beverly Grove; a rusting metal sign at Beverly Grove and Tower Grove points the way to "Bellavista." The small Spanish-style house at 9898 Beverly Grove is the former aviary; its distinctive circular roof can only be seen by driving up to the top of Beverly Grove and driving back down.

Across Benedict Canyon at 1436 Bella Drive is **Falcon Lair,** Rudolph Valentino's Spanish-style hideaway. When Valentino bought the house in 1925, his popularity was such that the high wall alone couldn't keep out the female fans, so the screen's first great Latin lover added floodlights, guards, and half a dozen dogs, including three Great Danes and two mastiffs, to keep his admirers at bay. Inside the house were floors of travertine marble, velvet drapes, and, on the wall of the entrance hall, a life-size portrait of Valentino in the guise of a Saracen. The Falcon Lair name can still be seen chiseled into the columns on both sides of the entrance gate, and the metal falcon flag still flies above the main tower on the far side of the courtyard.

One of the most easily seen star estates is **Owlwood** at 141 Carolwood Drive, just south of Sunset Boulevard in Holmby Hills. Through the tall iron-filigree gates is a great view of the broad lawns, imposing fountain, and neoclassical-style mansion built in 1934 by United Artists executive

Joseph Schenck and lived in since by the couples of Tony Curtis and Janet Leigh and Sonny and Cher.

The hilly area north of Sunset west of Beverly Glen Canyon could easily be mistaken for more of Beverly Hills, but it's actually **Bel Air,** a part of the city of Los Angeles. Developed in the twenties by oil millionaire Alphonzo Bell, Bel Air was designed to outclass Beverly Hills, and it did, with immense white gates at the main entrance, a private security patrol, and a rule that no movie people were allowed. That lasted until the Great Depression, when even movie stars' money became welcome. Once the gates were opened, the movie riffraff swarmed in and never left.

Bel Air is roughly centered around the **Bel Air Hotel** on Stone Canyon Road (see Hotels). It's rather luxurious now, but the bungalows were originally the Bel Air Estates sales office, and much of the hotel was originally the Bel Air Stables. The horses should have had it so good.

Two adjacent houses on Bellagio Road are recognizable for their film roles in which they portrayed average upper-middle-class houses, in spite of the fact that either would sell for well over a million dollars. The house at 10431 Bellagio was Dabney Coleman's house in *9 to 5*, where Jane Fonda, Dolly Parton, and Lily Tomlin tied up their no-good boss and hung him from the ceiling, while the house next door at 10445 Bellagio shows up occasionally in "Knots Landing."

A house that could *never* pass as an average upper-middle-class residence is the forty-room French chateau-style mansion at 750 Bel Air Road. Built by contractor Lyman Atkinson from 1935 to 1938 at a cost of $2 million as a surprise for his wife, the house came with a thirty-nine-by-seventy-foot living room, gold doorknobs, and a Baccarat crystal chandelier. In honor of Atkinson's successful investments in oil, a gold oil derrick served as the bathtub faucet in the master bath. Unfortunately for Atkinson, he was the one who got the surprise; his wife hated the house and refused to live in it. After leaving the house vacant for seven years, Atkinson sold it for $200,000, one-tenth of what it had cost to build. Today, the house is probably the most famous house in Bel Air, but not because of Atkinson's sad tale. The estate is where the **Beverly Hillbillies** lived when they moved to "the hills of Beverly . . . Hills, that is . . . swimmin' pools . . . movie stars."

Of course, there's more to Beverly Hills than just places where stars live, eat, and shop. The **Academy of Motion Picture Arts and Sciences,** home of the Academy Awards, is in Beverly Hills, at 8949 Wilshire Boulevard (213-278-8990). (See Attractions.)

South of the Golden Triangle, at the intersection of Olympic Boulevard and Beverly Drive, is one of the very few monuments to film and filmmakers in the Southland. This is the **Monument to the Motion Picture People Who Helped Save Beverly Hills.** A spiral of motion-picture film rises to a pinnacle above bas-relief figures of Will Rogers, Mary Pickford and Douglas Fairbanks, Valentino, Tom Mix, Harold Lloyd, Conrad Nagel, and director Frank Niblo. In 1923, the honorees, all residents of Beverly Hills, helped convince their neighbors not to go along with Los Angeles's plan to annex their small city.

The **Twentieth Century-Fox Studios,** 10201 West Pico Boulevard in Century City (213-277-2211), aren't technically in Beverly Hills, but they're just next door. Century City, a billion-dollar-plus development of offices, hotels, shopping centers, and condos, was in fact once the Fox back lot. When the Fox Film Corp., founded by former exhibitor William Fox, outgrew its studio in Hollywood, the company bought a parcel of land owned by the Janss Company, the developer of Westwood, and spent $10 million building the ninety-six-acre Movietone City, the first studio built from the ground up for sound. Fox Movietone News, the company's newsreel division, had already made history in 1927 with its sound newsreel of Charles Lindbergh's takeoff from Roosevelt Field, Long Island on the start of his trans-Atlantic journey, and Fox was one of the first film pioneers to see that the future lay in sound. William Fox's own future, however, lay in bankruptcy and jail after his attempt to build Fox into a super-conglomerate within the entertainment industry crashed along with the stock market in 1929. In 1935, Fox Film merged with Twentieth Century Pictures, a company originally founded by Joseph Schenck and Darryl Zanuck to produce films for United Artists. Twentieth Century-Fox's biggest stars in the thirties were Tyrone Power, Sonja Henie, Henry Fonda (in *The Grapes of Wrath*), and Betty Grable; today, the company's biggest stars are C3-PO, R2-D2, and Luke Skywalker and his *Star Wars* companions.

The Fox lot is the most impressive of all the studios. The original sound stages were built in the monolithic WPA style along a broad central avenue, yet tucked into corners of the remaining 68.24 acres of the lot are such gems of Hollywood dreamland as the Old Writers Building (no, it's not just for old writers), a Hansel-and-Gretel fantasy cottage topped with a weathervane in the shape of a quill pen and an inkwell. For a brief time in 1982, the studio was open to the public through a Gray Line tour, but the tour no longer operates. You can drive or walk part way onto the lot, however, and see the spectacular New York street, complete with its Victorian-style elevated station, built for *Hello, Dolly* in 1969. See it fast, though; only the depressed real-estate market is keeping what's left of the studio from being turned into condos by the new owner, oilman Marvin Davis.

Finally, we return to the residential flats of Beverly Hills for a look at another old-time movie studio that's now a house. The **Witch's House** of Beverly Hills (you'll know why it's called that as soon as you see it) was originally the administration building of the Irvin Willat Studios in Culver City. Willat had been a cameraman for Carl Laemmle's IMP Company and had shot a number of Mary Pickford films for director Thomas H. Ince before building his own Irvin Willat Productions in Culver City. Willat's art director, Harry Oliver, designed the office building to look like a witch's cottage; with its peaked roofs and tilted windows, any self-respecting witch would have been proud to call it home. The cottage was often used as a set in Willat's pictures. Willat Studios closed in the twenties when Willat joined Famous Players-Lasky, and in 1931 the building

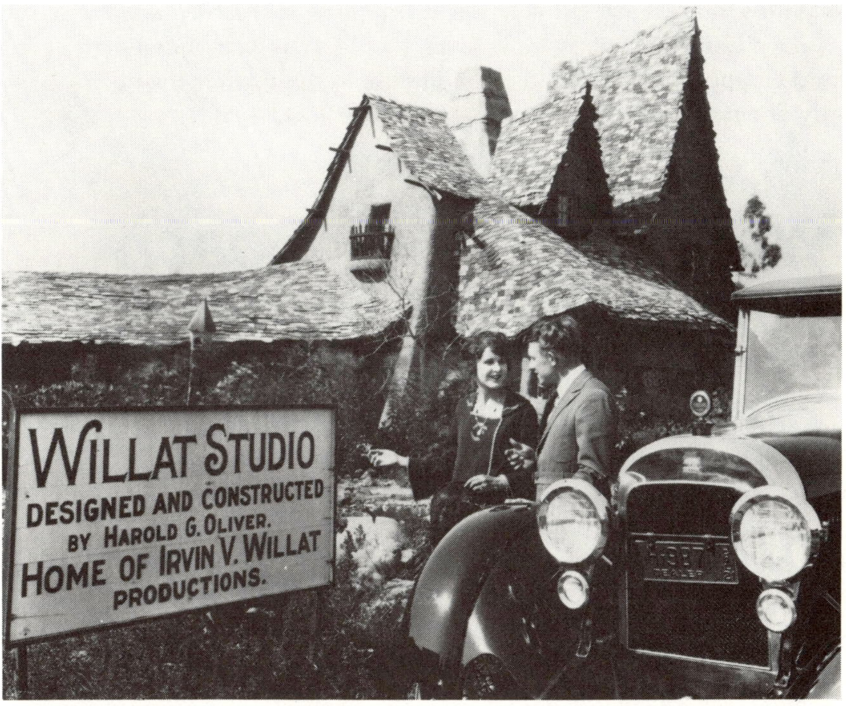

Once the headquarters building for the Willat Studios, whose specialty was B mystery and suspense films, this medieval European fantasy is now just another house in Beverly Hills.

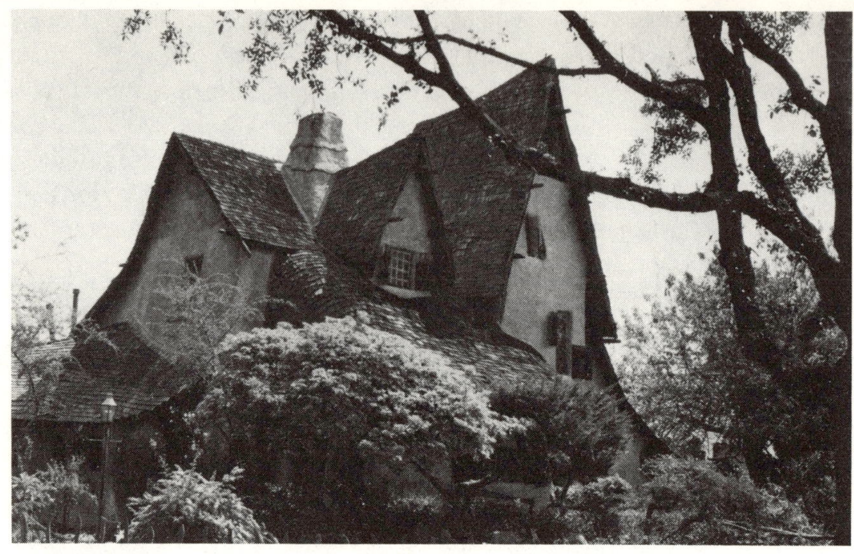

was moved to its present location at the corner of Carmelita Avenue and Walden Drive in Beverly Hills. Even then, its days as a movie set weren't over; it appeared as John Gielgud's house in the blacker-than-black comedy about the American way of death, *The Loved One.*

West Side and Beach

The west side of Los Angeles, including roughly everything west of the San Diego Freeway, plus Culver City, which is cut off from the heart of town by the Santa Monica Freeway, is the home of two important studios, a handful of important sites, and when Malibu is included, dozens of stars' homes.

Between 1908 and 1910, three major companies set up shop in Santa Monica: Kalem, Vitagraph, and a New York Motion Picture Co. unit under the direction of director Thomas H. Ince. While Vitagraph made do with a rented Pacific Electric Rail Road freight station, Ince's studio, modestly named Inceville, covered twenty thousand acres of Santa Ynez Canyon, where Sunset Boulevard now meets the sea. Besides director Francis Ford, brother of John Ford, Ince had a staff of seven hundred, numerous stages and other buildings, and a stock company of sorts in the

Miller Bros. 101 Ranch Show, a Wild West show Ince found wintering in Venice and enticed into moving to Inceville by offering to hire their horses and riders for his Kay Bee and Dommo Westerns. Ince's biggest hit wasn't a Western, though, but a 1916 epic in the DeMille mold, *Civilization.* The success of *Civilization* led Ince to abandon Inceville and build not one, but two studios in Culver City, both of which are still in business.

Ince's first Culver City studio was built by Ince for the Triangle Film Corp. on land donated by developer Harry Culver, who hoped to attract other industry using the studio as bait. Construction was finished in 1916, with the administration building facing Washington Boulevard behind a very long and very imposing row of classical columns. Ince left Triangle in 1918, and the following year Sam Goldwyn, late of Famous Players-Lasky, bought the lot with his partner, Ed Selwyn. In 1921 Goldwyn had a falling out with his company's board of directors and left, leaving his name behind. Three years later, the Goldwyn Pictures company—without Sam Goldwyn—needed help in completing the epic *Ben Hur,* and called in Louis B. Mayer to take over. Mayer was already involved with Marcus Loew's Metro Pictures, which also joined in the deal, creating **Metro-Goldwyn-Mayer Studios (MGM)** in 1924. MGM quickly became known as the "Home of the Stars," with everybody from Gable to Garbo under contract, and the studio's most famous films were lavish costume dramas such as *A Tale of Two Cities, The Wizard of Oz,* and *Doctor Zhivago.*

MGM was a leader in the rush by the studios to sell their birthright for a quick buck; much of its huge collection of props was sold at auction in the early seventies, about the time its huge back lot, where Gene Kelly went singin' in the rain, was demolished to make room for the Studio Estates residential development. The more compact studio continues to operate at full bore, however, especially since its recent merger with United Artists. The columns still stand along Washington Boulevard, but the main studio entrance is in front of the governmental-appearing Irving Thalberg Building at 10202 Washington Boulevard (213-558-5000).

With the settlement he got when Triangle Studios was sold to Goldwyn, Ince went down the road and built the Ince Studios, now the **Laird International Studios,** on fourteen acres at 9336 Washington Boulevard (213-836-5537). Construction began in 1919, and when the studio was finished in 1920, it had three glass-roofed stages and a swimming pool. Ince made a number of Westerns and other films on his new lot before he died in 1924.

Cecil B. DeMille then bought the studio for himself with financing from Producers Distributing Corp., using it to film the original version of *King of Kings* with the now-proverbial cast of thousands. Producers Distributing sold out to Pathé, but DeMille didn't like the deal, and he sold his own interest in the lot to Pathé as well.

A flurry of merger activity by banker Joseph P. Kennedy, the father of John, Robert, and Edward, turned Pathé into RKO, and RKO Radio Pictures threw the lot into full swing with *King Kong,* which reused the forty-foot high gates from DeMille's *King of Kings* as the gate to the native village. RKO's highest moment came in 1941 when it hired a young radio director to make his first feature film. The director was Orson Welles, and the film was *Citizen Kane.*

Meanwhile, in 1935, David O. Selznick quit working for his father-in-law, Louis B. Mayer, to go into business for himself. He started Selznick International Pictures and leased the studio from RKO. The distinctive main office building on Washington Boulevard, modeled by Ince after Mt. Vernon, became Selznick's trademark. Selznick made a number of memorable pictures, but none more memorable than his 1939 masterpiece, *Gone With the Wind.* Twelve fire companies stood by as Atlanta was burned on the studio's back lot; the fire marshals gave Selznick one hour to get it right.

Financial trouble forced Selznick out in 1949, and RKO, purchased by Howard Hughes a year before, took back the lot. Production dropped off, and the Culver City studio was shut down shortly after the filming of *The Thing* in 1951. In 1955, Hughes sold RKO to General Teleradio, a subsidiary of General Tire and Rubber, which in turn sold it to Lucille Ball and Desi Arnaz's Desilu a year later. Most of Desilu's shows, including "I Love Lucy," were shot at their Hollywood studios, and by 1963 the Culver City lot was just about empty, except for the George Stevens production of *The Greatest Story Ever Told* and the pilot for a new Desilu series, "Star Trek," which was filmed on Stage 15, the same stage that had once housed *Citizen Kane's* Xanadu.

Lucy bought out Desi's share in Desilu in 1960 and sold the company and both lots to Paramount in 1967. The Culver City lot was used, sparsely, as a rental lot; by 1970, the only show on the lot was the television series "Lassie." Paramount sold the lot to an investment company, which changed the name to the Culver City Studios and brought in the TWA Airline Film and TV Promotion division as a tenant. TWA installed a number of full-size airline interior mock-ups, which are seen in about 90 percent of all airline interior shots.

Finally, in 1979, the studio was sold once again, this time to a Canadian company, King's Point International, which changed the name to Laird International Studios. Laird continues to operate as a rental studio; one of

TWA's mock-ups on Stage 15 (the former *Citizen Kane* and "Star Trek" stage) was used in the spoof of all airplane movies, *Airplane,* and carrying on a tradition of excellence started by *Gone With the Wind* and *Citizen Kane,* the studio also provided a home for the Universal production, *E.T.*

Today, the lot is closed to the public, but the colonial-style administration building still stands on Washington and other buildings dating from the twenties can be seen around the corner of the lot on Ince Boulevard.

The studio remnant still standing on the west side is a small part of the **Vitagraph Studio** at 1438 2nd Street in Santa Monica. The abandoned brick building, the oldest masonry building in Santa Monica, was built in 1875 as the Rapp Saloon. In addition to being part of the Vitagraph Studio—the main studio building stood just to the north—the saloon also served as Santa Monica's first town hall.

The west side's studios may be closed to the public, but there are a pair of motion-picture locations open to the casual visitor, as well as one historic set that's open for dinner! All three are in Marina Del Rey, the county-owned marina that saw 172 days of filming in 1982. **Burton Chase Park** at the west end of Mindanao Way (213-823-4571) is a waterfront park with a fishing dock and picnic area that's used for scenes requiring a nautical backdrop in television shows such as "Charlie's Angels," "Fantasy Island," and "CHiPs," while **Fisherman's Village** at 13723 Fiji Way (213-823-5411), a row of Cape Cod–style shops and restaurants, has appeared in segments of "The Mod Squad," "The FBI," "Mission: Impossible," and "Trapper John, M.D." The park is open daily dawn–10:00 P.M., and most shops at Fisherman's Village are open daily 11:00 A.M.–9:00 P.M., with the restaurants open later.

If you'd like to dine in a set from one of the most famous musicals of all time, make reservations at Marina Del Rey's **Le Club** restaurant at 404 Washington Street (213-822-3939). Now a gourmet French restaurant, Le Club was originally a medieval-style restaurant called The Abbey, and the castlelike interior was built with sets from the film version of *Camelot* with Richard Harris.

Not long ago, newspapers made light of a search for a nude statue somewhere in a park in Boston, allegedly modeled by a young Bette Davis. The west side has a similar statue of its own, not of Bette Davis, but of a young **Myrna Loy.** Loy, the Warner Bros. star of the twenties, was a student at Venice High School, 13000 Venice Boulevard, when she posed for a statue of Victory on the school's grassy lawn. The head of the statue was blown off in 1979 by vandals, but was restored the following year. Immortality in Hollywood sometimes takes strange forms.

The west side and the beach areas have always been popular with stars, and there are a number of former residences that are standouts in the area.

Brentwood, which extends north and south of Sunset Boulevard west of the San Diego Freeway, has at least two former stars' homes of more than casual interest. A low, one-story Spanish-style house at 12305 5th Helena Drive south of Sunset is the house where **Marilyn Monroe** was found dead of an overdose of barbiturates in 1962, while the modest chateau-style house at 426 North Bristol Avenue, north of Sunset a little farther west, is the former residence of **Joan Crawford** and the house where most of the incidents described in Christina Crawford's *Mommie Dearest* took place. The walls around the houses held in more secrets than their neighbors knew.

Farther west, at 14253 Sunset, is a house with happier memories. This is **Will Rogers State Park** (213-454-8212), the former ranch of the only man in Hollywood who never met a man he didn't like. Born on a ranch in Oklahoma, Rogers first made his name as a trick roper with the Ziegfeld Follies before moving to Hollywood, where he appeared in twenty films and became the number-one box-office star in 1933.

Although he was the first honorary mayor of Beverly Hills, where he lived, Rogers built a small weekend cottage on his Pacific Palisades ranch in 1922. In 1928, he moved to the ranch full-time, expanding the cottage

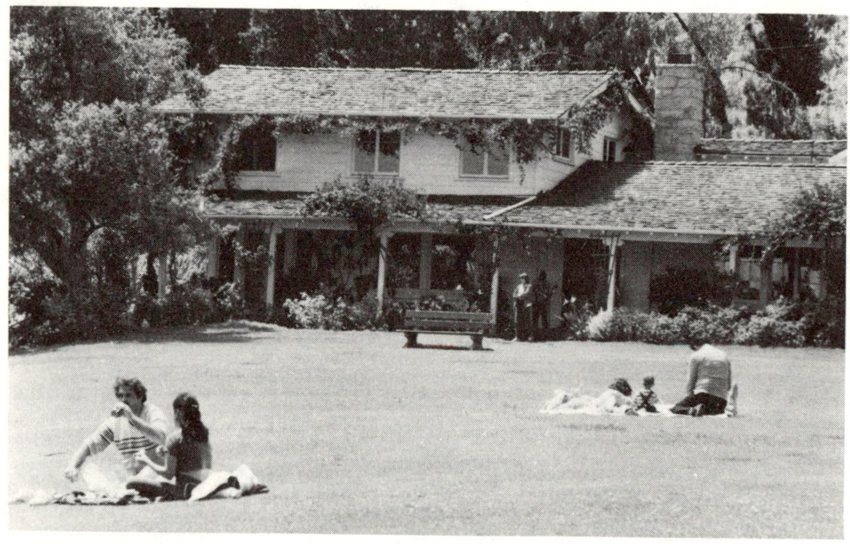

The Will Rogers estate in Pacific Palisades is one of the few movie-star mansions open to the public.

to its present thirty-one rooms. The house and grounds are maintained as they were when Rogers was alive. The living room is filled with Western-style furniture, Indian rugs, lariats, and other mementoes of Rogers's travels and career.

The 186-acre ranch includes miles of trails, stables, large picnic grounds, and a polo field that's still used by the Park West Polo Club. A visitors center feature's Rogers memorabilia and a continuous film of his life. The living room of the house is open, with a ranger on hand to answer questions, and you can take a self-guided audio tour of the rest of the ranch. The park is open daily (except Thanksgiving, Christmas, and New Year's Day) 8:00 A.M.–5:00 P.M.; the ranch house is open 10:00 A.M.–5:00 P.M. Admission is $2 per carload, or free to those on foot.

The beach at Santa Monica is the location of one famous movie location and a number of former Hollywood celebrities' beach "cottages," none of which could be bought today for less than a million dollars. The location site is, of course, the **Santa Monica Pier** at the western end of Colorado Avenue (213-393-5555). The pier, the oldest part of which was built in 1909, has a hodgepodge of beach-style shops, restaurants, and attractions, ranging from Doreena the Palm Reader to a handful of carnival midway games, a video-and-mechanical-game arcade, and a bumper-car ride. It's best known for its fully restored turn-of-the-century carousel, which figured prominently in the Paul Newman–Robert Redford film, *The Sting*. If you don't remember seeing a pier in that movie, don't worry; a matte shot made it look like the carousel was in old Chicago instead of on the beach. The pier's thirties ambience has also made it a popular location for such television shows as "The Gangster Chronicles," "Charlie's Angels," "Marcus Welby, M.D." and "CHiPS." The pier is open daily, and the carousel is usually open 10:00 A.M.–9:00 P.M. daily during the summer and weekends only during the winter.

The stretch of beach north of the pier was once known as the **Gold Coast;** this was Hollywood's first beach-resort community. You can see the houses as you drive along Pacific Coast Highway, or you can park in one of the beach lots and see them from the beach side. Unlike the Malibu Colony, this is a public beach, so no one will stop you from walking along it. The house at 707 Palisades Beach Road is the former home of Irving Thalberg, the boy genius who made MGM a major studio and who was the inspiration for the title character in F. Scott Fitzgerald's novel, *The Tycoon*. Next door at 705 Palisades Beach Road is the beachfront home of Douglas Fairbanks and Mary Pickford. Fairbanks kept the house after the divorce in 1936, and he died there in 1939. Louis B. Mayer's Spanish-style house at 625 was designed by MGM art director Cedric Gibbons, and the house was later owned by Peter and Patricia Lawford; her brothers, John, Robert, and Edward Kennedy, were frequent visitors. Jesse L. Lasky, whose company produced *The Squaw Man,* the first Western feature shot entirely in Hollywood, and who later merged with Adolph Zukor's company to form Paramount Pictures, built the house at 609 Palisades Beach Road. Lasky's partner Sam Goldwyn, who became the "G" in MGM after being forced out of Paramount by Zukor, built the house at 602 Palisades Beach Road; later on it was lived in by Mervyn LeRoy, who produced *The Wizard of Oz.* The house at 546 Palisades Beach Road was built by Darryl Zanuck, who cofounded Twentieth Century Pictures. The most modern looking house on the Gold Coast is the one at 514 Palisades Beach Road; it was designed in 1938 by Richard Neutra for Albert Lewin, the head of the story department at MGM who later produced *The Good Earth* and *Mutiny On the Bounty.* Anita Loos, the writer of *San Francisco* and *Gentlemen Prefer Blondes,* built the house at 506; funny man Harold Lloyd built 443; and the house at 237 Palisades Beach Road, was built by Louis B. Mayer, the "M" in MGM.

The greatest of all the Santa Monica beach mansions was mostly demolished in the fifties, but its remnants are still larger than any of the other houses on the Gold Coast. Located at 415 Palisades Beach Road is **The Sand and Sea Club,** a colonial-style private beach club that offers its 2,000 members such amenities as paddle tennis courts, a swimming pool, a bar and restaurant, massage facilities, 370 feet of roped-off beach, and forty-six private cabanas. The club buildings are the servants' quarters of Marion Davies' former home, which originally cost several of William Randolph Hearst's millions to build in 1928. The house had 750 feet of beach frontage, a ballroom, a theater, fifty-five bathrooms, thirty-seven antique fireplaces, and a basement bar imported from Britain that had originally been an Elizabethan pub. Nothing was too good for Hearst's mistress, or the parties she threw at the house; a lavish costume party with a thousand guests and a 125-piece orchestra was just an average Saturday night.

In 1946, Davies moved with Hearst to a house in Beverly Hills and sold the beach house, which became the Ocean House Hotel. When the hotel failed, the main house was torn down in 1956, leaving only the servants' wings, whose fate is currently up in the air. The city of Santa Monica has approved a plan that would allow The Sand and Sea Club to continue operating while becoming partially public by selling daily memberships

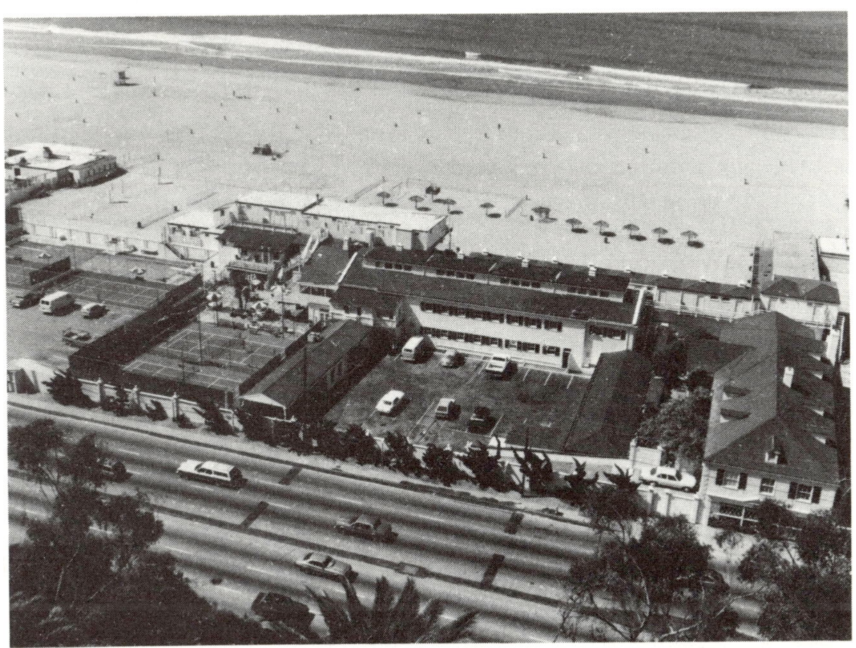

In Xanadu did Kublai Khan a stately pleasure dome decree. . . . The oceanfront Sand and Sea Club, an expensive private beach club in Santa Monica, is housed in the former servants' quarters of Marion Davies's beach house, built for the actress in 1928 by her lover, William Randolph Hearst.

and opening a restaurant. The State Parks and Recreation Department, which owns the property, has rejected the plan and seems determined to turn yet another remnant of Hollywood history into a parking lot. *Sic semper stupidity.*

A little farther north is a building that figured in one of Hollywood's enduring mysteries. The Spanish-style building at 17575 Pacific Coast Highway was once **Thelma Todd's Sidewalk Cafe,** a popular star hangout. One morning in 1935, Todd, the "Ice Cream Blonde" of Laurel and Hardy and Marx Brothers movies, was found dead of carbon monoxide poisoning in the garage of her apartment above the cafe at 17531 Posetano Road. The night before, Todd had been to a party given in her honor by Ida Lupino at the Cafe Trocadero on the Sunset Strip. Todd was found in her Lincoln, still dressed in her mink coat and jewels. But she hadn't driven home. She had been given a ride and dropped off at her cafe, and then walked the flight of 100 steps from the cafe to her apartment. Few believed it was suicide, especially those who talked about Todd's rumored connections with Lucky Luciano, gambling, and drugs. Today, the cafe is the home of Paulist Productions, the group of Catholic fathers who produce the television series, "Insight."

Malibu was a sleepy country town for decades, even after the founding of the Malibu Colony as a private retreat for stars who liked to mix their beachfront living with privacy. Today, it's still a hard-core suburban neighborhood, except that your neighbor is likely to be Johnny Carson, Barbra Streisand, or Olivia Newton-John, and gossip over morning coffee at the Colony Coffee Shop is as likely to revolve around the running feud between next-door neighbors Larry Hagman and Burgess Meredith as it is what kind of traps to set for snails. Other residents range from Bob Dylan to the boy guru Maharaj-Ji. If you want to see stars candid and close up, nearly any restaurant, dry cleaners, or supermarket in Malibu will do (see Stargazing), but Malibu has a few other movie-related attractions as well.

The **Malibu Pier,** at 23000 Pacific Coast Highway (213-456-8030) in what passes for downtown Malibu, appears in the background of almost every movie or television show shot in Malibu. **Alice's Restaurant,** at the foot of the pier (213-456-6646), has nothing to do with the movie of the same name, but it *is* the place where Alan Alda and Jane Fonda debated the relative merits of Los Angeles and New York in Neil Simon's *California Suite.* **Paradise Cove,** a privately owned (but open to the public) beach at 28128 Pacific Coast Highway (213-457-2511), is where private investigator Jim Rockford parked his trailer in the James Garner series, "The Rockford Files." Admission to the Malibu Pier is free; admission to Paradise Cove, open daily from sunup to sundown, costs $5 per carload, unless you're going to the Sandcastle Restaurant in the cove.

San Fernando Valley

When Los Angeles was first settled, the San Fernando Valley, the vast, flat area on the north side of the Hollywood Hills, was pretty much deserted. It was too dry and too out-of-the-way to be much good for anything. Then in 1913, water came to the Valley via the newly completed California Aquaduct, and farms took the place of scrub. The following year, the movies moved over the hill from Hollywood and settled in for good. Today, most of the farms are gone—"The Real McCoys," whose television farm was set in the Valley, wouldn't recognize the place—but the movie studios, movie ranches, and other related sites are still there to interest the fan.

The most important sites in the Valley are the five studios and one movie ranch located almost next to each other in the Burbank–Studio City area. The first—and still the largest—is **Universal Studios** at 100 Universal City Plaza on Lankershim Boulevard in Universal City (818-508-2515). Today, Universal is most famous for its tour (see Chapter 1), but the tour is just the latest entry in the history of the seventy-year-old studio.

Like many of the early film magnates, Carl Laemmle, Universal's founder, started out as an exhibitor, opening his first nickelodeon in Chicago in 1905. In 1909, Laemmle became one of the first exhibitors to publicly announce that he would no longer pay a license fee to the Motion Picture Patents Co., the trust that controlled all legal filmmaking in the United States. He set up his own production company, Independent Motion Pictures, IMP for short.

In 1912, Laemmle changed the name of his company to the Universal Film Manufacturing Co., and, in an attempt to outrun the trust's goon squads, moved first to Cuba, then to Los Angeles, where he bought David Horsley's Nestor Film Company, the old Blondeau Tavern in Hollywood. Universal quickly outgrew the small studio, and in 1914 Laemmle bought a 230-acre chicken ranch in the San Fernando Valley. A year later, in a ceremony attended by Buffalo Bill Cody and Thomas A. Edison, Laemmle opened the new Universal Studios. Besides having an 80-by-410-foot stage that could accommodate a dozen film companies at once and all the necessary shops and laboratories, the studio had a sheep pasture, a zoo, and plenty of leftover chickens, whose eggs Laemmle sold to the surrounding residents.

Universal's early films included John Ford's first feature, *Straight Shooting;* Erich von Stroheim's first picture, *Blind Husbands;* and Lon Chaney's greatest film, *The Phantom of the Opera.* The opera house built for the picture still stands inside Stage 28. For a decade after the *Phantom,* Universal was the prime home of the horror film, unleashing *Dracula, Frankenstein, The Mummy,* and *The Invisible Man* upon a ready-to-be-frightened world. Other studio output ranged from dozens of bread-and-butter Westerns to *All Quiet on the Western Front,* which won the studio's first Oscar for Best Picture in 1939.

Comedies and genre films predominated in the forties. Universal stars included Abbott and Costello, W. C. Fields, and Basil Rathbone, who, as Sherlock Holmes, stalked Moriarty and other sinister villains through Universal's back lot. Universal's merger with International Pictures in

1946 didn't change much; for every *Hamlet* with Laurence Olivier, the studio released dozens of films with Ma and Pa Kettle, the now-infamous Bonzo, and Francis the Talking Mule.

In 1958, the Music Corporation of America (MCA), which had started as a talent agency in 1925, bought the studio lot, but not Universal Pictures, which remained on its old lot as a tenant. Then in 1962 MCA bought Universal Pictures as well. Universal productions since that time have included the blockbusters *Jaws, The Sting, National Lampoon's Animal House, Smokey and the Bandit, American Graffiti,* and *E.T.*, the all-time box office king, in addition to television productions too numerous to contemplate, much less mention.

Today, Universal's 420 acres includes the studio and tour, the 6,200-seat Universal Amphitheater, a hospital, three banks, three restaurants—Victoria Station, Womphopper's Wagon Works, and Fung Lum—and two hotels, the brand new Sheraton Towers and the Sheraton-Universal. The restaurants and hotels are open daily; see Chapter 1 for tour information.

The First National Exhibitors Circuit was formed by twenty-seven major exhibitors in 1917 in an attempt to break out from under Paramount's thumb; the new company's first two stars were Charlie Chaplin, who signed for $1 million for eight films, and Mary Pickford. Both stars had their own studios, though, so it wasn't until 1922 that First National built its own giant lot in Burbank. Today it's known as **The Burbank Studios (TBS)**, the home of Warner Bros., which bought First National in 1928, and Columbia Pictures, which moved onto the lot in a joint real-estate venture with Warner Bros. in 1972. Warner Bros. pictures shot on the lot can be identified by decade. The thirties saw James Cagney in *Public Enemy* and Edward G. Robinson in *Little Caesar,* along with *42nd Street* and the rest of the Busby Berkeley dance classics. The forties was the decade of Bogart, who appeared in *High Sierra, The Maltese Falcon, The Treasure of the Sierra Madre, The Big Sleep,* and, of course, *Casablanca,* while the fifties brought James Dean and *East of Eden, Rebel Without a Cause,* and *Giant.* In the sixties, Warner Bros. was the home of big-budget musicals, *My Fair Lady, The Music Man,* and *Camelot,* and of *Bonnie and Clyde* and *The Wild Bunch.* Since then, Warner Bros.' biggest hits have included *The Exorcist* and *Superman,* while Columbia's biggest film since it moved onto the lot has been *Close Encounters of the Third Kind.* Other tenants of The Burbank Studios include the Ladd Company, Rastar Productions (makers of *Annie*), and Clint Eastwood's Malpaso Productions. The Burbank Studios VIP Tour (see Chapter 1) is the best tour in town for seeing how a major movie studio really works. The studio is at 4000 Burbank Boulevard in Burbank (213-954-6000).

About two blocks north of the main TBS lot at 3701 West Oak Street is the **TBS Ranch,** a forty-acre back lot built by Columbia in 1935 and known as the Columbia Ranch until the creation of TBS. Presently the ranch's main full-time tenant is "Fantasy Island"; this is where Mr. Rourke's cottage, the Fantasy Island hotel, and most of the sets used in the guests' different fantasies are located. The ranch is off-limits to the public, but you can walk in far enough to get a look at the ranch's colonial-street set.

Here's an easy one: on what studio's lot can you find streets with names like Minnie Mouse Lane and Dopey Drive? When Walt Disney moved to Los Angeles from Kansas City in 1923 with dreams of making cartoons, he started out by making an animation stand in the garage in back of his Uncle Robert's house on Kingswell Avenue in Hollywood. Today, the **Walt Disney Studios** at 500 South Buena Vista Street in Burbank (818-840-1000) has seventy-seven buildings spread over forty-four acres. Disney's first black-and-white cartoons were a series called Alice Comedies, which put a live girl—Alice, of course—in the middle of an animated cartoon. The first "Alices" were made in the rear of a real-estate office down the street from Disney's uncle's house, which Walt and his brother Roy rented for ten dollars a month. They soon moved to the storefront next door, which became the first Disney Brothers Studio, and in 1926, the brothers moved into a new studio they had built on nearby Hyperion Avenue. The huge success of Mickey Mouse, born in 1928 in the first sync-sound cartoon, *Steamboat Willie,* forced the expansion of the Hyperion studio, which grew from 1,600 square feet to 20,000 square feet in 1933. By 1938, the studio had outgrown the buildings on Hyperion altogether and slopped over into two apartment houses and a brace of rented offices in Hollywood. With the profits from their first feature-length animated film, *Snow White and the Seven Dwarfs* (1937), the Disney brothers bought fifty-one acres in Burbank, moving to the new lot at the start of 1940. Three buildings made the move as well: the mail-room bungalow, the personnel building (formerly the animation-trainees building), and the Shorts Building, which now houses the wardrobe, music, and makeup departments.

There are only four sound stages on the small lot. Stage 1 was one of the original 1940 buildings and was first used for some of the live-action scenes in *Fantasia.* Stage 2 was built in 1949 with the help of Jack Webb, who used it for filming his television series, "Dragnet." Stage 3 was built in 1953–54 to house the water tank needed for underwater and special-effects filming for *20,000 Leagues Under the Sea,* and Stage 4 was completed in 1958 and first used for *Darby O'Gill and the Little People.* The small back lot contains the "Zorro" set that served as both the Pueblo de Los Angeles and the city of Monterey in that series, as well as a Western street, a 1920s midwestern residential street, a town square, and an area of pools and caves used for jungle scenes. The electric and machine shop on the back lot has its own place in Disney history; this is where many of the vehicles used at Disneyland were built, including two of the full-size steam locomotives.

The studio is generally closed to the public, but the Walt Disney Archives collection of historical material is open by advance arrangement to qualified students and writers. For more information, see Libraries.

NBC Studios, at 3000 West Alameda Street in "beautiful downtown Burbank" (818-840-4444), was built in 1952 and became the West Coast headquarters of the network with the demolition of the NBC Studios in Hollywood in 1964. A wide variety of game shows, local news broadcasts, and special productions, including many of the Bob Hope specials, are made here, but NBC Studios is best known as the home of "The Tonight Show Starring Johnny Carson." You can take a guided tour ofthe studios (see Chapter 1), or, if you're in a hurry, you can look through the VIP gate between the buildings n Olive Street and see Johnny Carson's car (the last parking space on the right) and the metal awnings of the infamous NBC Commissary.

Studio City got its name from the studio that's presently known as **CBS/ Fox Studio** at 4024 Radford Avenue (213-760-5000). The studio was built by comedy producer Mack Sennett in 1928 and was leased by Mascot Pictures in 1933. The Indian Village, South Sea Village, jungle, and Western Street on the back lot at that time provided the flavor of Mascot's serials. In 1935, Mascot merged with Consolidated Film Laboratories and Monogram to form Republic Pictures, and the lot became Republic Studios. Monogram withdrew in 1936, but Republic stayed alive, churning out one B picture after another, including a ton of John Wayne, Gene Autry, and Roy Rogers Westerns. The studio turned out an occasional A picture as well, including Orson Welles's *Macbeth* and John Wayne's *The Sands of Iwo Jima* and *The Quiet Man,* plus cult classics such as Nicholas Ray's *Johnny Guitar* and the last successful serial, *King of the Rocket Men.* Republic went under in 1959 after television had wiped out the need for B pictures, and the studio became a rental lot with Dick Powell's Four Star International as its first tenant. CBS leased the lot in 1963 and bought it in 1967, using it for such series as "Gunsmoke" and "The Mary Tyler Moore Show." In 1982, Twentieth Century-Fox bought a half-interest in the lot as part of its plan to tear down its own studio in Century City.

Parts of the back lot are still standing and can be seen by walking north on Colfax Street from Ventura Boulevard at the east end of the studio.

Few bona fide movie stars made their homes in the San Fernando Valley; it was considered the sticks, certainly not high class enough for most celebrities. Two who bucked the trend were **Clark Gable** and **Carole Lombard,** who built their ranch house in the Valley at 4543 Tara Drive in Encino, a short distance south of Ventura Boulevard. Today the ranch has been subdivided into the Clark Gable Ranch Estates, but their blue-and-white frame-and-stone ranch house is visible through the iron gate, or by driving around to the rear of the house on Ashley.

Malibu Creek State Park on Las Virgenes-Malibu Canyon Road, about midway between the Ventura Freeway and the ocean in Agoura, is made up of the former Bob Hope Ranch, the former Ronald Reagan Ranch, and the former Twentieth Century-Fox Ranch that was used for lavish productions like John Ford's *How Green Was My Valley, Planet of the Apes,* and *Tora! Tora! Tora!* and is still in use for film and television productions, including the late, lamented "M.A.S.H." in which it doubled for Korea. These are the hills the choppers would fly over at the start of every episode; the "M.A.S.H.," set itself was located at Crags Road and Lost Cabin Trail. Also in the park is Century Lake where Butch Cassidy and the Sundance Kid jumped off the cliff to escape the Pinkerton posse. Fifteen miles of hiking trails crisscross the six-thousand-acre park, which straddles Malibu Creek. The park is open daily except during fire closures from 8:00 A.M. to dusk; a visitors center staffed by docents is open from noon to 4:00 P.M. Sunday. Parking costs $2 per car on weekends and holidays, but foot traffic is always free. For information on fire closures, call 213-454-2372 during the fire season; for information on the park itself, call 213-706-1310 during the week or 818-880-4089 on weekends.

The west end of the San Fernando Valley is the location of a number of movie ranches, two of which are open to the public. The 336-acre **Paramount Ranch,** on Cornell Road two miles south of the Ventura Freeway at Kanan Road in Agoura, is a remnant of the forty-thousand-acre ranch where Paramount filmed such epics as *The Adventures of Marco Polo* (1937), which featured elephants and two thousand decorated horses. The old Western set, used in "The Virginian," "Have Gun, Will Travel," "The Rifleman," and dozens of other television series and films, is still standing and still used from time to time, although most of the buildings and false fronts look like they could collapse at any moment. It's a great place for kids; it isn't as well preserved as the Western streets at Knott's Berry Farm or

Universal Studios, but here kids—and adults—can run through swinging saloon doors and jump off balconies to their hearts' content. The ranch, which is also the location of the annual Renaissance Pleasure Faire, is part of the Santa Monica Mountains National Recreation Area, and a ranger gives guided nature walks on Sundays year-round and tours of the Western set on weekends in the summer. The ranch is open daily from dawn to dusk. Call 818-888-3770 for more information.

At the north end of the Valley are still more movie ranches, most of which, unfortunately, are closed to the public. The most historic is the **D. W. Griffith Ranch,** where troops of the Ku Klux Klan once thundered across the hills in Griffith's first epic, the 1915 *The Birth of a Nation.* The ranch, at the end of Arroyo Street north of Foothill Boulevard in San Fernando, is now owned by developer Fritz B. Burns, who turned part of the flatlands into a business park. An historical marker sits on the corner of Foothill and Vaughn Street.

Farther west, in Chatsworth, is the **Hope Ranch,** part of the ten thousand or so acres in the San Fernando Valley owned by Bob Hope. Built by Western star Ray "Crash" Corrigan, the ranch, once called Corriganville, used to be open to the public, who could watch staged movie fights and other stunt demonstrations and wander through Fort Apache, used in "The Adventures of Rin Tin Tin" television series. The ranch, located on Iverson Lane just off old Santa Susana Pass Road, is now off limits to the public.

North of Chatsworth in the Newhall-Valencia area are two other movie ranches that are closed to the public, Gene Autry's **Melody Ranch** on Oak Creek Canyon and Walt Disney Studios' **Golden Oak Ranch** on Placerita Canyon Road, and one old-time star's estate, the **William S. Hart Ranch** at 24151 Newhall Avenue (805-259-0855), that's open to the public. Hart, the number-one star of silent Westerns, bought the ranch (then called the Horseshoe Ranch) in the early twenties. The 250-acre ranch, now a county park, includes the original 1910 ranch house, now a museum filled with photographs, posters, paintings, ten-gallon hats, and other Hart relics, all fenced off from sticky fingers by metal mesh grates. On the hill behind the ranch house are the dogs' cemetery and a combination bunkhouse-game room, and at the top of the hill is another museum, a Spanish-style residence, La Loma de Los Vientos (the Hill of the Winds), built by Hart in 1925–28 and filled with Remingtons, Russells, and other Western artworks. A barnyard area next to the ranch house has sheep, goats, horses, reindeer, and plenty of fowl, including chickens running loose, while a pen on the other side of the hill is home to a number of buffalo, first given to the park by Walt Disney in 1962. The park is open free of charge daily 10:00 A.M.–7:00 P.M. in summer and 10:00 A.M.–6:00 P.M. in winter, and La Loma de Los Vientos is open 10:30 A.M.–3:00 P.M. Wednesday through Friday and 10:30 A.M.–4:30 P.M. Saturday and Sunday. Guided tours of the house are given every half hour.

Two sites in the Newhall area saw a lot of filming in the early days. **Beale's Cut,** a fifty-foot-deep, ten-foot-wide pass cut through a hillside, was created in 1859 by General Edward Beale and the men of nearby Fort Tejon; sixty years later, Tom Mix jumped his horse over it and hordes of movie cowboys rode through it. There's a parking area on the east side of Highway 14, about a mile north of the junction with Highway 5, and Beale's Cut is about a quarter-mile walk northeast.

A few miles farther up Highway 14 is **Placerita Canyon State Park** at 19152 West Placerita Canyon Road (805-259-7721), the site of California's first gold rush, which started in March 1842 when herdsman Francisco Lopez pulled up some wild onions and found traces of gold on the roots. In the teens, cinematic herdsmen sought another kind of gold here. The park is open daily dawn to dusk.

Farther east along Highway 14 in Agua Dulce is the most spectacular Western—and science fiction—movie site of all, the 745-acre **Vasquez Rocks County Park** at 10700 West Escondido Road (805-268-0991). The high, jagged rock formations are the most visible feature of the San Andreas earthquake fault. The area was the home of the Aliklik Indians, a branch of the Shoshone, from 200 B.C. into the 1700s, and later served as a hideout for Southern California's most notorious bandit, Tiburcio Vasquez. Vasquez was betrayed by one of his lieutenants (after sleeping with the lieutenant's wife) and was wounded in a shootout in the rocks, escaping only to be captured in a cabin in the Hollywood foothills. More recently, the area has served as the backdrop for both classic motion pictures, such as *The Charge of the Light Brigade,* and classic television series: the rocks doubled for nearly every alien planet visited by Captain Kirk and Mr. Spock in "Star Trek." The park is open daily from 8:00 A.M. to a half hour before dark.

Cemeteries

The glamour of Hollywood doesn't necessarily end at the grave. Of all the famous final resting places, **Forest Lawn Memorial Park** in Glendale generally is conceded to be the last word in star quality. After all, it was the inspiration for the book, *The Loved One,* a merry look at the lives and loves of mausoleum workers.

More than three hundred acres of green rolling hills are dotted with marble and bronze artworks. In the Great Mausoleum, a sepulchral voice welcomes visitors to a giant stained-glass re-creation of Leonardo da Vinci's *Last Supper.* Elsewhere, there is *The Crucifixion,* said to be America's largest religious painting; a copy of Italy's famed *Paradise Doors;* and an exhibit of every coin mentioned in the Bible. Two churches, the Church of the Recessional and the Wee Kirk O' the Heather, are available for weddings, christenings, and funerals. When Jean Harlow's funeral services were held at the Wee Kirk O' the Heather in 1932, Jeanette MacDonald and Nelson Eddy sang at the service, and Clark Gable was a pallbearer. Gable returned to Forest Lawn's churches twice more; in 1942, Carole Lombard's services were held at the Church of the Recessional, and in 1960, Gable's own services were held there as well.

Forest Lawn officials like to emphasize the brighter side of this eternal resting place. There are no guides to the locations of famous inhabitants, but friendly groundskeepers may point out a few, such as the interment spots of Errol Flynn and Spencer Tracy near the Freedom Mausoleum.

A list of the famous buried at Forest Lawn reads like a "Who's Who" of who's departed.

At Forest Lawn in Glendale, 1712 South Glendale Avenue (213-254-3131), are Humphrey Bogart, Alan Ladd, Clark Gable, Carole Lombard, Jean Harlow, Clara Bow, Rex Bell, W. C. Fields, Errol Flynn, Marie Dressler, Theda Bara, Robert Taylor, Spencer Tracy, Dick Powell, Joe E. Brown, Gracie Allen, Harold Lloyd, Nat King Cole, Jeanette Macdonald, Walt Disney, Chico and Gummo Marx, Francis X. Bushman, Russ Colombo, Jean Hersholt, theater owners Sid Grauman and Alexander Pantages, and producers Irving Thalberg and David O. Selznick.

The filmland famous are also laid to rest in several other locales, open, like Forest Lawn, free to the public.

Forest Lawn in Hollywood Hills, 6300 Forest Lawn Drive (818-984-1711), has Freddie Prinze, Stan Laurel, Charles Laughton, Ernie Kovacs, Buster Keaton, Dan Duryea, Clyde Beatty, Horace Heidt, and Rex Ingram.

At **Hillside Memorial Park,** 6001 Centinela Avenue, Culver City (213-776-1931), a large bronze bust of Al Jolson, visible from the nearby San Diego Freeway, overlooks a tinkling fountain. Also interred here are Jack Benny, Eddie Cantor, Jeff Chandler, David Janssen, and Vic Morrow.

Dating from the 1890s, **Hollywood Memorial Park Cemetery,** 6000 Santa Monica Boulevard (213-469-1181), is the final resting place of a number of early stars, including Rudolph Valentino, "The Sheik." Also here are Peter Finch, Douglas Fairbanks, Sr., Nelson Eddy, Tyrone Power, Paul Muni, Marion Davies, Cecil B. DeMille, Adolphe Menjou, Clifton Webb, Norma and Constance Talmadge, Jesse Lasky, Harry Cohn, Cass Daley, Virginia Rappe, silent star Barbara LaMarr, Peter Lorre, Victor Fleming, and the eldest of Charlie Chaplin's children, Charlie Chaplin, Jr.

At **Holy Cross Cemetery,** 5835 West Slauson Avenue, Culver City (213-670-7697), Bela Lugosi is said to be laid to rest swathed in his trademark cape. Also here are Sharon Tate, Bing Crosby, John Ford, Mario Lanza, Zasu Pitts, Frank Lovejoy, Rosalind Russell, Joan Davis, Edgar Kennedy, Spike Jones, Joe Flynn, Kid Ory, Charles Boyer, Jimmy Durante, Louella Parsons, William Lundigan, Mack Sennett, and Barney Oldfield.

Just off busy Wilshire Boulevard is **Westwood Memorial Park,** 1218 Glendon, Westwood Village (213-474-1579), the final resting place of legendary sex symbol Marilyn Monroe. Officials will point out her crypt. After some twenty years, ex-husband Joe DiMaggio cancelled a three-times-a-week standing order of a half-dozen roses for Marilyn, but fans still visit the spot daily, usually gathering for ceremonies on the anniversaries of her death and her birth. Natalie Wood is buried here in a quiet green space, sheltered by nearby camphor and redwood trees. Also interred here are Richard Conte, Jay Flippen and Dorothy Stratten, Oscar Levant, Minnie Riperton, and Stan Kenton.

Man's best friends are not forgotten. In Calabasas, the **Los Angeles Memorial Pet Park** has served as a final home for dearly beloved animals since 1929. Interred at the park are Petey, star of the Our Gang comedies, at least one of the Lassies, Hopalong Cassidy's horse Topper, Rudolph Valentino's dog Kabar, and pets belonging to such celebrities as Humphrey Bogart, Gloria Swanson, Freddie Prinze, Sally Struthers, Bob Newhart, and Lionel Barrymore. Officials say that star owners are rarely in evidence, but ordinary pet owners bring flowers, monuments, and even Christmas trees in fond memoriam.

4

THE GUIDED TOURS

Most visitors to Hollywood who don't have much time to spend and who want to see stars end up taking a guided tour. Not only do those visitors rarely see stars, but they also have to take the information given out by most tour guides with a grain of salt. While some tours, notably those given by historical groups, are excellent, most commercial tour guides are left to their own devices when it comes to acquiring the information they pass on, and the level of knowledge—and the degree of accuracy—can vary widely from one guide to the next.

Gray Line Tour #2, which covers Hollywood and Beverly Hills, is probably the best-known tour in town. The tour takes about four hours, including brief stops at the Hollywood Bowl and the Chinese Theater and an hour-long lunch or dinner break at Farmers' Market. While Gray Line's drivers usually pride themselves on being well informed, on one typical tour, the driver wrongly identified Echo Park as the location site for "McHale's Navy" and "Gilligan's Island" (both were filmed on studio back lots); pointed out a mansion on Sunset Boulevard as Shirley Temple's childhood home (she never lived there for a day); and said that interior shots for "CHiPS" were filmed at the California Highway Patrol office overlooking the Hollywood Freeway (they were shot on the MGM lot in Culver City). To be fair, those were almost the only real mistakes the driver made, although there were additional sins of omission, such as not pointing out Sunset-Gower Studios, the former home of Columbia Pictures for fifty years. For the most part, the guide provided a steady stream of informative and accurate commentary on the history of the places his bus was passing.

The big disappointment with the Gray Line tour in the past was the stars' homes portion; the company's fifty-three passenger buses were too large to negotiate the narrow streets of Bel Air and the hilly portion of Beverly Hills. When the city of Beverly Hills banned tour buses from its residential streets altogether, the Gray Line buses moved further west to Brentwood Park in the city of Los Angeles, but then L.A. put a weight restriction on the buses, banning them from residential streets there, also. Now Gray Line uses vans that fall under the weight limit, so Brentwood Park and Bel Air are back on the tour, although much of Beverly Hills' residential area is still off-limits. As for the homes themselves, while most of those in Bel Air qualify as mansions, the ones in Brentwood aren't any bigger than upper-middle-class houses in Peoria. As far as stars themselves are concerned, you'll be lucky to see their cars in the driveways.

The tour leaves daily from the Gray Line depot at 1207 West Third Street in downtown Los Angeles at 1:30 P.M.. Shuttle buses run between the depot and most major hotels, but travel time can take up to an hour and a half in each direction. The fare is $15.00 for ages twelve and over, $7.25 for ages five to eleven, and free for children under five. For more information, call 213-481-2121.

The Avis to Gray Line's Hertz is Starline Tours, which operates from a booth in the forecourt of the Chinese Theater. **StarLine Tour #1** is their stars' homes tour, and it has its good and bad points when compared to Gray Line's tour. The StarLine tour takes just two hours, since it doesn't go to the Hollywood Bowl or Farmers' Market, and because it starts and ends at the Chinese Theater in the heart of Hollywood, you also save travel time to and from downtown Los Angeles.

On the debit side, StarLine's drivers seem to put less of a premium on accuracy than on making the tour interesting. The stars' addresses StarLine's drivers point out are usually correct, as are their mention of homes that were used as locations in different films, but much of the other information they pass on is straight from Fantasyland.

StarLine's Tour #1 operates daily every half hour 10:00 A.M.–6:00 P.M. June 15 to September 15 (every 15 minutes when things get really busy) and 10:00 A.M.–4:00 P.M. September 16 through June 14. The fare is $11.00 for ages twelve and over, $5.50 for ages five to eleven, and free for children under five. The fare includes a shuttle bus from your hotel to the StarLine terminal at the Chinese Theater, 6845 Hollywood Boulevard, and back. For more information, call 213-463-3131.

Hollywood Heritage, the organization that restored the *Squaw Man* barn and turned it into the Hollywood Studio Museum (see Attractions), publishes a self-guided walking tour of Hollywood Boulevard in association with the Los Angeles Conservancy and occasionally gives guided tours of Hollywood neighborhoods and locations, such as Whitley Heights and the old Hollywood Athletic Club (now the Berwin Entertainment Complex). The tours concentrate on general history, but since it's impossible to talk about the history of Hollywood without mentioning the history of motion pictures, they're always of interest to visitors intrigued by the golden days of the movies. You can pick up the Hollywood Boulevard tour map for $3 at Hollywood Heritage's office at 1350 North Highland or at the Hollywood Studio Museum across from the Hollywood Bowl. For information on the group's other tours, call 213-466-6782 or 874-BARN.

The Broadway Theater District Tour, operated by the Los Angeles Conservancy, is for those who are as interested in where the movies are shown as in where they're made. The Broadway Theater District, which stretches from Third Street to Ninth Street on Broadway in downtown Los Angeles, is listed on the National Register of Historic Places as the home of eleven movie palaces of the twenties. The two-hour walking tour includes the interiors of three of the most spectacular theaters and the exteriors of all but one of the rest.

The tour, which starts every Saturday at 10:00 A.M. in the lobby of the Subway Terminal Building at 417 South Hill Street, costs $5 and reservations must be made in advance by calling the Los Angeles Conservancy at 213-623-CITY.

Finally, for a view of Hollywood and Beverly Hills unlike any other, **Red Baron Scenic Air Tours** offers a ninety-minute tour of Southern California in a seven-passenger single-engine aiplane. The tour includes much more than Hollywood—the 350-kilometer route goes all the way down to Newport Beach—but you still get incomparable views of a number of motion-picture studios and all the swimming pools in Beverly Hills. Tours leave from Burbank Airport, and the $65 per person cost includes hotel pickup. There's a minimum of three persons. For foreign visitors, tours are narrated in French, German, Spanish, Italian, Dutch, Japanese, and Chinese.

Red Baron Scenic Air Tours is based at San Fernando Airport, 12910 Foothill Boulevard in the San Fernando Valley. Call 213-896-6667 for information.

5

STAR-GAZING

Imagine window-shopping with Frank Sinatra, working out next to Jane Fonda, or munching on some Chinese chicken salad alongside Mick Jagger.

It may sound like a fan's fantasy, but seeing the stars in Hollywoodland is not only possible, at times it's downright unavoidable. After all, the film, television, and music industries are still highly concentrated in the Los Angeles area, and celebrities have to eat, too.

Stars also have to work, play, get their hair done, and, especially in Beverly Hills, they have to shop.

Saturday afternoon on Beverly Hills' plush **Rodeo Drive** can resemble a slow-motion Rolls-Royce rally, with Jaguars, Mercedes, and Porsches filling in the gaps. The beautiful people are out to browse, and the Rodeo area, which falls between Wilshire and Little Santa Monica boulevards, is a great place to catch them in action. Here's a sampling of stores often patronized by celebs.

Giorgio, 273 North Rodeo Drive (213-278-7312), a chic shop that sells men's and women's designer clothes, shoes, and accessories, from little silk blouses to $4,000 ball gowns. There are walls full of celebrity photos. Customers have included Loretta Swit, Jimmy Stewart, Barbara Walters, and Jack Lemmon. There's a small coffee and liquor bar for shoppers who need a boost (customers and guests only, please), a comfy wingback chair near a fireplace for the weary, and a pool table for the bored. They think of everything.

At **Gucci,** 347 North Rodeo Drive (213-278-3451), status stripes are everywhere at status prices. Don't despair. Cutesy Gucci matches start at $5 or so, and tie bars can be had for a mere $10. Bigger spenders might be interested in something a little more substantial, such as the Gucci-model Cadillac Seville. It has Gucci-trademark-fabric top trim, plus tasteful trademark wheels, and, of course, a *G* hood ornament, all for about $39,000.

The more thrift-minded may prefer **Gucci on Seven,** 211 North Canon Drive (213-278-3451). This store is devoted to discount Gucci items. The name derives from the fact that the New York Gucci store's discount goods are on the seventh floor.

A star simply must be well shod, and dozens march right into **Right Bank Shoe Co.,** 301 North Rodeo Drive (213-275-4559) to pick up a pair or two (especially pricey Maud Frizon labels), plus bags and accessories. Sometime shoppers include Tatum O'Neal, Cher, Donna Summer, Joan Collins, Jaclyn Smith, and Olivia Newton-John.

DiFiori Shoes, 363 North Camden Drive (213-274-4352), is not far away. Their own designer-label shoes and boots often adorn the fabulous feet of Sissy Spacek, Goldie Hawn, and Candice Bergen.

Back on Rodeo Drive, one of the new additions is a multi-level marble marvel, a mini-shopping mall known as **The Rodeo Collection,** 421 North Rodeo Drive (213-276-9000). Vuitton, Fendi, Ungaro, Nina Ricci, Yves Saint Laurent, and Gianni Versace are among the labels represented. The marble, brass, and greenery alone are worth a visit. Even the garage, with a waterfall-effect fountain and topiary trees, has star quality. Upstairs, the **Excelsior** restaurant (213-275-5100) features French and Italian cuisine, with dinner entrees at about $20 and up. Downstairs, the **Pastel** restaurant offers Continental cuisine with a prix fixe dinner at about $25 per person, and less expensive lunches. Olivia Newton-John, Ursula Andress, and several foreign film stars have been spotted in Pastel's pastel surroundings.

Just across the way from The Rodeo Collection is **Bijan,** 420 North Rodeo Drive (213-273-6544). The store sells menswear and special gift items. Bijan gained publicity by selling a fourteen-carat-gold inlaid Colt revolver at $10,000 a shot. That does include a mink glove. Shoppers who still need a security blanket may prefer a king-size bedspread at about $48,000 in mink, or $98,000 in chinchilla. The store is open by appointment only. Customers have included well-dressed stars George Hamilton, Bob Hope, Jack Lemmon, and Warren Beatty.

To soak up some Southern California mood, there's **Camp Beverly Hills,** 9640 South Santa Monica Boulevard (213-274-8317). The boutique sells casual wear in rainbow colors, many pieces emblazoned with the store's palm-tree logo. One need never grow up in Beverly Hills, if one chooses not.

Another major shopping area, several miles to the east, is the great tan

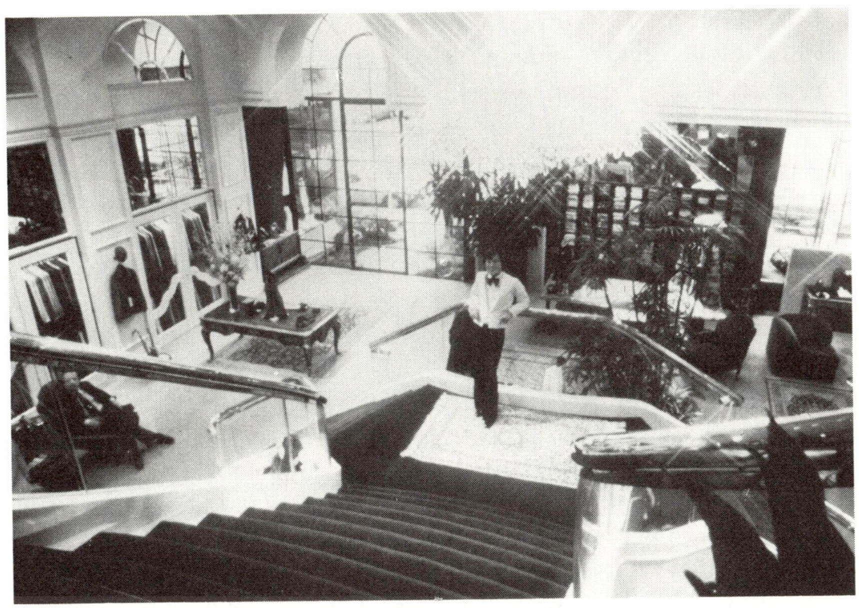

Bijan on Rodeo Drive, the most expensive shop on Beverly Hills' most expensive street. Need a $300 shirt or a $10,000 gold-plated .38 caliber revolver? By appointment only, if you please.

monolith, **The Beverly Center,** 8500 Beverly Boulevard (213-850-0070). The Beverly, a giant mall with marble interiors, houses boutiques and restaurants, as well as two major department stores, Bullock's and the Broadway. Several stars, including Bette Midler and Angie Dickinson, have been spotted browsing at the Broadway; others have been seen dining at the store's third floor deli and restaurant. Snacking shoppers may include Glenda Jackson, Valerie Perrine, and Bob Hope. Then there's the **Hard Rock Cafe,** a trendy dining spot in the mall that is detailed later in the restaurant section.

Meanwhile, take a jaunt over to **Maxfield** (also known as Maxfield Bleu), 9091 Santa Monica Boulevard (213-275-7007). The store, which specializes in high-fashion designer wear, modestly avows that "everybody" shops there, especially those in the music world. But they prefer to name no names, so you'll have to see for yourself.

The Melrose Avenue area is full of trendy shops that attract an eclectic mix of new wave and old guard.

Let it Rock, 7310 Melrose Avenue (213-934-4416), is filled with what is described as English rock-and-roll clothes. Customers sometimes include Timothy Hutton, and members of the Rod Stewart band.

Koala Blue, 7366 Melrose Avenue (213-651-5506) is owned by singer-actress Olivia Newton-John and her former singing partner, Pat Farrar. The shop features Australian-designed clothes for children and adults. Olivia herself helps design some items. There are also an Aussie milk bar for milkshaes and meat pie snacks, and a giant video screen to showcase Australian bands. Celebs such as Dudley Moore, Susan Anton, and Barbara Streisand have been seen scooping up "bear" essentials.

Melon's, 8739 Melrose Avenue, and **Monon's Shoes,** 8750 Melrose (213-854-7734), display trendy designer wear in a totally "tubular" space of skylights, white tile, and plants. Bernadette Peters is among the stars who've been sighted ogling the merchandise, but the store keeps its rock-music clients very hush, hush.

There's no secret about who's riding high in a certain Western wear shop. The **Edward H. Bohlin, Inc.** saddlery shop, 172 Verdugo Avenue in Burbank (818-846-2851), has been around for decades, although it moved from its old Hollywood location in 1983. Burt Reynolds had them work up a little ol' belt buckle—in gold, of course—with his initials in pavé diamonds. The shop makes buckles, belts, and other Western gear. Besides Clint Eastwood and former Presidents Eisenhower and Johnson, customers have included a cavalcade of cowboy stars: Gene Autry, John Wayne, Gary Cooper, and Tom Mix. Superstar Mae West, always alert to male interests, also patronized Bohlin's. The accommodating firm once made Miss West a sterling-silver telephone, presumably so that admirers could call up and see her sometimes.

Ms. West's blonde tresses were one of her trademarks. Even today, a star's best friend may be her hairdresser. With shear good luck, you may be able to see the stars at these shops, most in Beverly Hills.

Allen Edwards, 455 North Rodeo Drive (274-8575), is where Donna Mills, Cathy Lee Crosby, Sally Field, and Raquel Welch are often spotted.

At **Maurice Jose,** 9426 Little Santa Monica Boulevard (213-274-4562), co-owner Jose (Joe-say) Eber has trimmed the locks of such luminaries as Cher, Goldie Hawn, Farrah Fawcett, Angie Dickinson, Victoria Principal, Linda Gray, and Valerie Perrine. Eber reportedly charges $100 for a cut, but he's no longer taking new customers. Busy, busy.

The customers of **Menage A Trois,** 8822 Burton Way (213-278-4430), include Morgan Fairchild, Donna Reed, and Barbara Rush.

Co-owner John Isaacs of **Michaeljohn,** 414 North Camden Drive (213-278-8333), has done the hair for *Victor-Victoria,* among other films. A number of celebrities pop in, including Julie Andrews and Pamela Bellwood.

Milka Harp, 308 North Rodeo Drive (213-273-3773), has an eclectic clientele that has included little Drew Barrymore, Elizabeth Taylor, Barbara Walters, Swoosie Kurtz, and a number of Playboy playmates.

Strugar Hair Co., 455 North Doheny (213-273-1672), has a roster of star clients that has included Grace Slick, Jodie Foster, Yoko Ono, Tina Louise, and Cheryl Ladd.

All that shopping and primping can build up an appetite. There are hundreds of restaurants in Los Angeles, and scores of places have a Hollywood connection of some sort. Some are longtime star havens, some are celebrity owned, and others are simply favored by the ever-changing trendy set. We won't even attempt to include every hot spot, or to rate the food. But we will try to give you an idea of what to expect in a selection of restaurants. As for price, $ indicates low cost, $10 or under per person; $$ means moderate cost, $20 or under; and $$$ means more—often much, much more. Remember, lunches are usually cheaper, so even a $$$ place may be affordable for lunch. Operating days vary and many places insist on reservations, so always call ahead.

First, a few of what might be considered the "A list" of stargazing restaurants.

Ma Maison, 8368 Melrose Avenue, West Hollywood (213-655-1991), is *the* place for seeing stars. At one time, the restaurant even boasted an unlisted telephone number. Orson Welles has a standing reservation here when he's in town. Virtually everybody who's somebody (or would like to be) shows up. To name just a small cross section: Jack Lemmon, Michael York, Chad Everett, Jacqueline Bisset. $$$

Chasen's, 9039 Beverly Boulevard, Beverly Hills (213-271-2168), is home to old-line Hollywood. This bistro was founded by ex-vaudevillian Dave Chasen back in 1937. At one time, it offered a steam room and showers for the comfort of traveling entertainers who dined there. Chasen's famous chili is a favorite of President Reagan and Elizabeth Taylor. The restaurant is frequented by directors, writers, and established stars like Frank Sinatra and Jimmy Stewart. $$$.

Mr. Chow's, 344 North Camden Drive, Beverly Hills (213-278-9911), serves Peking cuisine and is said to be a favorite of Susan Anton, Dudley Moore, Bruce Dern, Jill St. John, Kenny Rogers, and, at times, the entire cast of the television series "Dynasty." $$$.

The northern Italian dishes at **Dan Tana's,** 9071 Santa Monica Boulevard, Los Angeles (213-275-9444), attract scads of stars like Cheryl Ladd, Carol Lawrence, Pia Zadora, Robert Stack, Gene Barry, and Paul Michael Glaser. $$$.

Le Dome, 8720 Sunset Boulevard, Los Angeles (213-659-6919), is a haven for rock and rollers, especially the well-to-do. After all, Elton John and Rod Stewart were among the original investors. The cuisine is French and Continental, and stars like Olivia Newton-John, Donna Summer, and Alice Cooper do stop by. $$$.

Out in Malibu, where many stars live, there are several places where the famous flock.

There's **La Scala** Malibu, 3835 Cross Creek Road (213-456-1979), which is not to be confused with **La Scala** in Beverly Hills, 9455 Little Santa Monica Boulevard (213-275-0579), or the **La Scala Boutique,** 475 North Beverly Boulevard (213-550-8288). All locations attract celebrities, but the Malibu restaurant is especially popular. Barbra Streisand, Larry Hagman, Lynda Carter, Johnny Carson, Eddie Albert, Billy Joel, Katharine Ross, and Telly Savalas are often seen there. $$$.

Not too far away is **Beau Rivage,** 96025 Pacific Coast Highway (213-456-5733). This is a romantic French spot where Larry Hagman, Olivia Newton-John, and other stars may be seen soaking up the atmosphere. $$

Then, there's **Holiday House,** 27400 Pacific Coast Highway (213-457-3641). The Continental cuisine and fabulous view of the ocean attract a star-studded crowd, including Olivia Newton-John, Barbra Streisand, Robert Redford, David Letterman, Christie Brinkley, Michael Landon, Mick Fleetwood, David Frost, and on and on. $$.

And, for deli fans, there's the **Malibu Gourmet,** 28925 Pacific Coast Highway (213-457-5802). You can take out, or eat in alongside Walter Matthau, Ali MacGraw, and Jan Michael Vincent. $.

Back in town, there's **Scandia,** 9040 Sunset Boulevard (213-278-3555). Considered by many one of Los Angeles's finest restaurants, Scandia serves Danish and Continental food to dozens of stars. $$$.

The **Los Angeles Palm,** 9001 Santa Monica Boulevard (213-550-8811), is a steak-and-lobster house based on the New York Palm restaurant. Caricatures of famous customers line the walls. You're almost sure to see someone notable, such as Elliott Gould, George Hamilton, Cheryl Ladd, Johnny Carson, or Steve Garvey. $$$.

If you're in a New York mood but want to spend a bit less, head for **Joe Allen's,** 8706 West Third Street (213-274-7144). It's got good American food, an often crowded bar, some actors, and, of course, transplanted New Yorkers. $$.

There are always new restaurants opening to attract Los Angeles trendsetters. Topping the list of current favorites is **Spago,** 8795 Sunset Boulevard (213-652-4025). Chef Wolfgang Puck defected from star-studded Ma

Maison to create his own restaurant here. Celebrities and would-be stars are flocking to sample such exotic fare as duck sausage pizza and pumpkin ravioli. For weekend reservations, you may need to call as much as three weeks in advance. Star clients include Donald Sutherland, Zsa Zsa Gabor, Peter Falk, Nell Carter, and Warren Beatty. $$$

Puck's new restaurant, **Chinois,** at 2709 Main Street, Santa Monica (213-392-9025 or 392-3037) is a nouveau-Chinese fantasy that's smaller and even harder to get into than Spago, but spiced with just as many stars. $$$.

Then there's **Trumps,** 8764 Melrose Avenue, Los Angeles (213-855-1480), an airy spot that's helping to bring back the tradition of afternoon tea, complete with scones, cucumber sandwiches, and other delicacies. Dinner is also served. Celebrity diners mix with artsy types from the nearby Pacific Design Center. $$$.

At **Bono,** 8478 Melrose Avenue, Los Angeles (213-651-1842), entertainer Sonny Bono has cooked up a restaurant popular with his friends and fans. The cuisine, of course, is Italian—plenty of pasta and fresh seafood, plus steak and veal for those to whom meat is a must. Sonny himself is there just about every night, and often supervises things during the day as well. Many celebs, including Elizabeth Taylor and Roger Moore, drop in, as does ex-wife Cher. Why not let Sonny do the cooking? $$.

Hard Rock Cafe, 8614 Beverly Boulevard in The Beverly Center (213-276-7605), is an all-American hot spot that may be the trendiest yet. It has an airy atmosphere, with wood floors and bars, green-and-white-tiled kitchen, and waitresses in old-fashioned white uniforms and caps. New wave and nouveau people line up, eagerly awaiting the cafe's ribs, chicken, burgers, fries (delicious), and such. $$.

And, for beach-bound trendsetters, there's the **West Beach Cafe,** 60 North Venice Boulevard, Venice (213-399-9246). California cuisine is served here in a light atmosphere, with artworks displayed on white walls. Or you can just have a drink and "people watch" at the bar. $$.

Just about every studio has a couple of nearby gathering places where stars, technical people, and friends hang out.

Near Paramount Studios:

Lucy's El Adobe, 5536 Melrose Avenue (213-462-9421), is a casual Mexican-style restaurant just across the street from the studio. Stars' photos line the walls. Noted customers include former California Governor Jerry Brown and Linda Ronstadt. $.

Oblath's, 723 North Bronson (213-465-1662), has been operating since 1923. The Mexican and Continental cuisine includes such specialties as deep-fried burritos and guacamole omelets. Burt Reynolds, Goldie Hawn, and other stars frequent the place. In fact, Paramount has an extension phone at the restaurant—a hot line to call actors needed on the set. $.

Nickodell, 5511 Melrose Avenue (213-469-2181), serves good Continental cuisine at reasonable prices. Studio stars drop in for a meal, or a drink at the bar. $.

Assistance League of Southern California Tearoom, 1370 North St. Andrews Place (213-469-1973), is not too far from Paramount. The restaurant is run by a charitable organization, and was once near several movie studios on nearby Sunset Boulevard. The studios folded, and starlets no longer wait tables as a publicity stunt as they did in the twenties and thirties, but the food is tasty and reasonably priced and studio personnel still drop in. Ted Knight often eats here between tapings, and Elizabeth Taylor has been known to stop by. It's okay if you look at the stars, but don't press, please. $.

Over by the Warner Hollywood Studios, there are more star-studded wateringholes:

Ports, 7205 Santa Monica Boulevard (213-874-6294), is a bar and restaurant with international cuisine and a notable Chinese chicken salad. This spot attracts all kinds of stars, such as Dick Van Dyke, Burt Reynolds, and Mick Jagger. $$.

The Cantonese and American food at **Formosa Cafe,** 7156 Santa Monica Boulevard (213-850-9050), once attracted such legendary figures as Clark Gable, John Wayne, and Jack Benny. A wide range of celebrities, including Buddy Ebsen and Robert Goulet, still sample the menu. $.

Studio Grill, 7321 Santa Monica Boulevard (213-874-9202), is a favorite spot for fanciers of such delicacies as duck pasta and lime and ginger shrimp. $$.

In Burbank, near NBC studios, you might try:

Chadney's, 3000 West Olive (818-843-5333), has American food for down-to-earth diners like Merlin Olsen, Charlton Heston, and Bob Hope. $$

Hamptons, 4301 Riverside Drive (818-845-3009), is where Paul Newman's salad dressing made its West Coast debut. Newman is a customer, as are Bo and John Derek, moguls Ray Stark and David Begelman, and other Hollywood folk who crave a simple burger, soup, or salad. $.

Near Universal Studios, there's:

Maison Gerard, 4100 Cahuenga Boulevard (818-766-3841), is called Mason's Garage by irreverent diners. Whatever the name, it's a favorite of stunt and technical people. $$.

Some celebrities, perhaps not content with existing restaurants, have come up with their own.

One of the best known is the **Ginger Man,** 369 North Bedford Drive, Beverly Hills (213-273-7585). Owners Carroll O'Connor and Patrick O'Neal offer saloon-style dining, with burgers and fresh fish the specialties. Celebrities like Johnny Carson, Mary Martin, Linda Evans, Alan Alda, Jack Lemmon, and, of course, the owners, often stop in. $$.

Ah Fong's, 424 North Beverly Drive, Beverly Hills (213-276-1034), is a Cantonese restaurant owned by character actor Benson Fong. Tons of celebrities, including Dustin Hoffman, Jon Voight, Lucille Ball, Nick Nolte, Jim Backus, Neil Diamond, Julie Andrews, and many more, stop by here. $$.

And if you're wondering where director Steven Spielberg hangs out when he wants a home-style dinner, try his mom's place, **The Milky Way,** 9108 West Pico (213-859-0004). Spielberg's mother, Leah Adler, says her dairy restaurant specializes in Kosher gourmet cooking, and under rabbinic supervision yet. There's no place like home. $$.

The **DDL Food Show,** 244 North Beverly Drive, Beverly Hills (213-859-2700) is owned by Dino De Laurentiis, the producer of such epics as the remake of *King Kong.* Ever the showman, De Laurentiis has made the interior of his restaurant-deli-take out and catering establishment look something like the inside of a sound stage, with all the food bathed in theatrical lighting. The cuisine is Italian, as you'd guess, and features too much of everything—40 kinds of pasta, 300 kinds of cheese. Since nothing succeeds like excess, the DDL Foodshow is sure to be a hit. $$.

If you're on a really tight budget, but still want to soak up the Hollywood scene, there's more:

The star-studded **Nate 'n Al's,** 414 North Beverly Drive, Beverly Hills (213-274-0101), is a favorite deli of actors, writers, and studio execs. Even the regulars have to wait in line to get in. Doris Day and Walter Matthau are just a couple of the stars who have the deli habit. $.

And for dessert, there's always room for **Hägen-Dazs,** 362 North Canon Drive, Beverly Hills (213-276-2086). If their luscious ice cream is rich enough for "Dallas" rich girl Charlene Tilton, it's probably good enough for you. $.

Now that you've had your fill of food, perhaps you'd like a change of pace? Some sustenance for the soul? Actors who want a change from their work on the big (or little) screen sometimes appear onstage in productions at smaller theaters.

The **Matrix,** 7657 Melrose Avenue (213-852-1445), is one of a number of small, legitimate theaters in the Melrose area. Paul Michael Glaser and Ian McShane are just two of the actors who've appeared at this one.

Pan Andreas, 8325 Santa Monica Boulevard (213-654-8780), is a repertory theater that has showcased the talents of Marcia Wallace, Dennis Christopher, and Dee Wallace.

Westwood Playhouse, 10886 Le Conte Avenue, Westwood (213-208-5454), has seen a number of stars, including Kate Jackson, Vincent Price, Gabe Kaplan, and Pat Carroll, trod the boards.

But maybe you'd rather spend the evening looking over some superstars of tomorrow, while at the same time keeping an eye out for the stars of today. Here's a sampling of Los Angeles nightlife. Remember, most clubs have a cover charge and/or a drink minimum.

Comedy Store, 8433 Sunset Boulevard (213-656-6225), is where such comics as Richard Pryor and David Letterman tried out their early material. New comedians are showcased seven nights a week in this Sunset Strip club, and established stars often come in to check out the old neighborhood.

Improvisation, 8162 Melrose Avenue (213-651-2583), is the Comedy Store's major rival. The Improv gave early exposure to such talents as Robin Williams, Steve Martin, and Betty Thomas. They still drop by, as do old-timers like the Smothers Brothers and Andy Kaufman.

An up-and-coming music club is the **Club Lingerie,** 6507 Sunset Boulevard (213-466-8557). The interior was the scene of the Bus Boys' set in the film *48 Hours.* All kinds of music are played here, from reggae and rhythm and blues to new wave. Famous listeners include Charlene Tilton, Britt Ekland, Lesley Anne Downs, and Mick Jagger.

For jazz buffs, there's **Mulberry Street,** 12067 Ventura Place, Studio City (213-980-8405). Serving homemade pasta, pizza, and other goodies, the club has an all-jazz format. On Tuesday nights, the Beverly Hills Unlisted Jazz Band holds forth, with George Segal sitting in when he's in town.

Back on Sunset Strip, there is the **Rainbow Grill,** 9015 Sunset Boulevard (213-278-4232), which is still a hangout for a number of rock stars.

Roxy, 9009 Sunset Boulevard (213-276-2222), features all kinds of rock and roll performed nightly in art deco ambience. This spot attracts celebrities as diverse as Dan Ackroyd and Rod Stewart.

For something completely different, there's **La Cage Aux Folles,** 643 North La Cienaga (213-657-1091). You're guaranteed to see top female stars—sort of. This club features female impersonators portraying such celebrities as Judy Garland and Dolly Parton. But real celebs stop in, too. Even legendary Lena Horne pops in for a look.

Enough of dissipation? Ready for a star-studded workout? The quest for fitness verges on fanaticism in Lotusland. Looking good may literally mean money in the bank. So, here's a brief sampling of health and racquet clubs where there's a good chance of spotting stars.

Jane Fonda's Workout, 369 South Robertson Boulevard, Beverly Hills (213-652-9464), is where fabulously fit Jane herself sometimes takes classes. Classes are often taught by young actresses, and the famous Fonda name attracts all sorts of international television and film crews doing stories on her workouts. A number of stars exercise at the Beverly Hills location, and regulars say the classes are strenuous. Sessions are available on a per-class or longer-term basis.

Richard Simmons' Anatomy Asylum, 9036 Santa Monica Boulevard, Beverly Hills (213-550-8879), is where Simmons, one of America's favorite exercise gurus, sometimes teaches. It's a membership club, but out-of-towners can attend on a per-class basis if they have proper identification to prove they don't live in Los Angeles. There are other Asylums in the Los Angeles area.

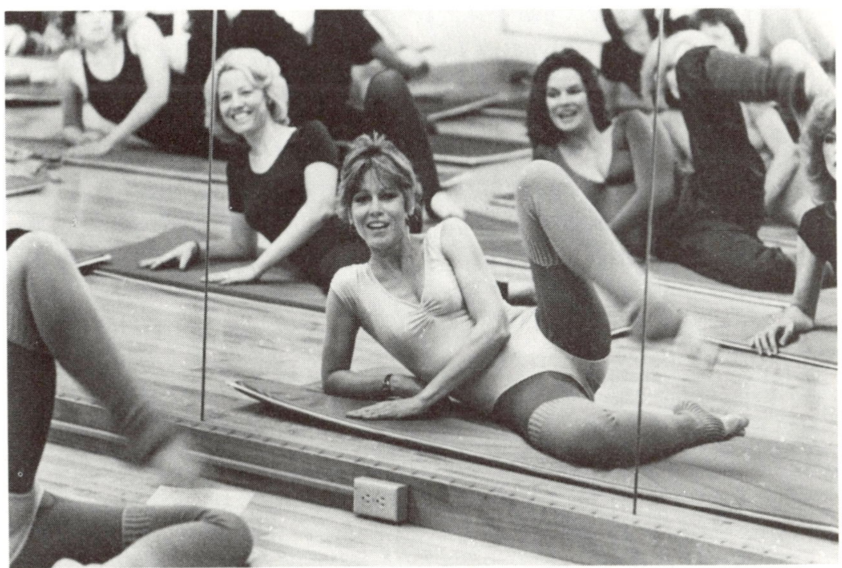

Many of the physical fitness salons in Los Angeles cater to celebrities, but Jane Fonda was the first celeb to turn the tables when she opened Jane Fonda's Workout in Beverly Hills.

Body Design by Gilda, 2080 Century Park East, Century City (213-553-2512), is a penthouse exercise studio that emphasizes the total body. Gilda Marx has shaped a number of celebs, and sometimes leads classes here. Fees for classes are on a yearly or per-class basis.

Century West Health Club, 2040 Avenue of the Stars, Century City (213-556-3312), is a full-service club in the ABC Entertainment Center that attracts lots of industry stars and execs. Membership only, but visitors staying at certain hotels, including the Beverly Hills Hotel, Beverly Wilshire, and Century Plaza, may use the facilities on a temporary-fee basis.

Ron Fletcher Company, 8330 West Third Street, Los Angeles (213-278-4777), offers individual and small-group instruction. Fletcher has worked one-on-one to shape many stars; Barbara Streisand, Raquel Welch, Ali MacGraw, Katharine Ross, and Candice Bergen are among the stars who've stayed sleek with his method.

Gold's Gym, 360 Hampton Street, Venice (213-392-3005), is a coed gym available for use on a daily or long-term basis. Muscle-man Arnold Schwarzenegger pumps iron here. Other fitness buffs include Jane Fonda, Morgan Fairchild, Linda Gray, Stefanie Powers, and Robert Wagner.

At **Robert Carreiro Physical Fitness Studio,** 722 North La Cienega Boulevard (213-652-3060), you can swing along with the stars on the trapeze, hanging rings, or parallel bars. Adult gymnastics and floor exercises are offered on a series or per-lesson basis. Jamie Lyn Bauer ("Bare Essence") and Kathleen Beller ("Dynasty") are just a couple of the stars who've worked out here.

For racquet-sport fans, there are several tennis and racquet clubs that allow visitor play or lessons.

Racquet Center, 10933 Ventura Boulevard, North Hollywood (818-760-2303), has tennis, paddle tennis, and racquetball courts open to the public by fee. Star players here include Fabian, Ricky Schroeder, and Gary Coleman.

Tennis Place, 5880 Third Street, Los Angeles (213-931-1715), has lessons and courts available to nonmembers. Often seen are Sidney Poitier, Gene Hackman, and Stacy Keach.

Toluca Lake Tennis Club, 6711 Forest Lawn Drive, Hollywood (213-851-6000), offers lessons to nonmembers, but they cannot play here. Studio execs frequent the courts, as do Gil Gerard and other celebrities.

If you prefer an indoor sport, how about some target practice at the **Beverly Hills Gun Club,** 12306 Exposition, West Los Angeles (213-826-6411)? The club offers a shooting range and instruction on a membership or fee basis. Sylvester Stallone is a partner in the business. Robert Culp,

Angie Dickinson, and Clint Eastwood are just a few of the many stars who've taken aim here.

If all else fails, be on the lookout for guaranteed star-studded occasions, such as premieres, charity functions, and special events held throughout the year:

The American Heart Association, 2405 West Eighth Street, Los Angeles (213-385-4231), holds the annual **"Hart to Heart" Celebrity Softball Game,** usually in January or February. Robert Wagner and Stefanie Powers, stars of the television series "Hart to Hart," lead teams to benefit the heart fund.

The **Glen Campbell Los Angeles Open** is a celebrity-filled golf tourney sponsored by the Junior Chamber of Commerce, 4004 South Bixel Street (213-482-1311). For $10 a day, you can check out the strokes of celebrated golfers like ex-President Jerry Ford, Bob Hope, Clint Eastwood, James Garner, and Jack Lemmon. The tourney is held each winter, usually at the Riviera Country Club, 1250 Capri Drive., Pacific Palisades (213-454-6591).

Finally, there's the annual **Hollywood Christmas Parade,** when television and film stars come out to ride with Santa down Sunset and Hollywood boulevards. The parade is usually held the Sunday after Thanksgiving, with plenty of stars old and new in attendance. The event is co-sponsored by the Hollywood Chamber of Commerce, 6290 Sunset Boulevard (213-469-8311), and singing cowboy Gene Autry reportedly was inspired to write "Here Comes Santa Claus" while riding in this annual extravaganza. Tradition!

6

STAYING WITH THE STARS: HOTELS

Only in Hollywoodland could a famous visitor dine tête-â-tête with his man-eating plant in one of the city's finest hotels without arousing comment.

And where else could a hotel guest keep a pet panther with the blessing of the management?

For that matter, where else could a Japanese emperor expect to be greeted by a mariachi band?

Los Angeles's hotels are a home away from home for some of the world's most famous people. Kings and presidents are regular visitors. Stage and screen stars work and play here, and some of that glamour seems to rub off on their temporary surroundings.

So here's a brief look at some of the city's most colorful, historic hotels, plus a quick glimpse at a couple of new stopping places building legends of their own.

First and foremost is **The Beverly Hills Hotel,** 9641 Sunset Boulevard (213-276-2251), which is also known as the Pink Palace and has a mystique thicker than Los Angeles smog. It was here that W. C. Fields once entertained his man-eating plant and that Jacqueline Susann wrote *Once Is Not Enough* and other novels. Reclusive billionaire Howard Hughes stayed here off and on for thirty years, sometimes renting multiple cottages and suites by the year. Aristotle Onassis spent most of World War II ensconced here. Modern movie mogul Marvin Davis is a permanent resident in a

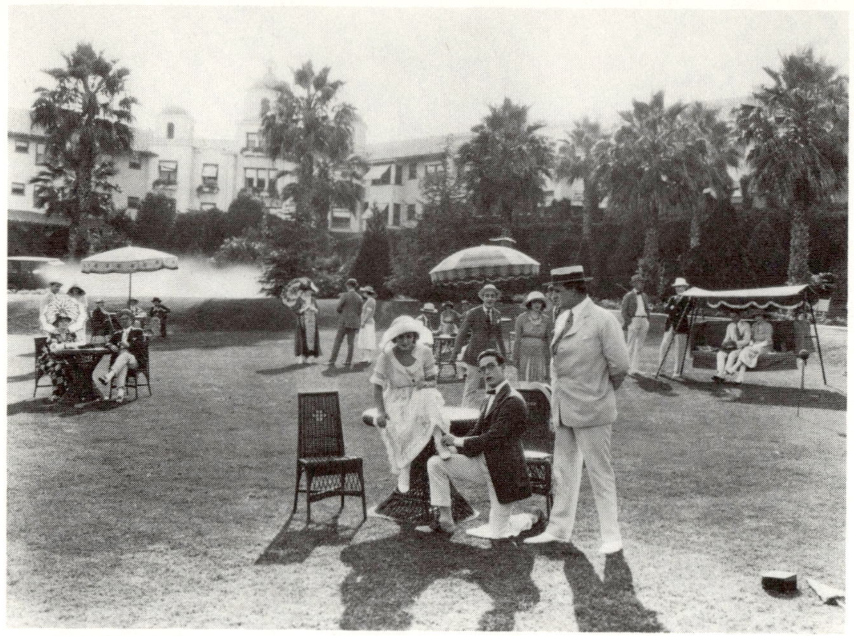

The Beverly Hills Hotel doesn't just house stars; it's been a star itself. Here Harold Lloyd films A Sailor Made Man *in 1921 on the hotel's lawn.*

bungalow decorated to his specifications. The tab is said to be some $300,000 a year. He must like it, and why not?

The sprawling hotel and grounds exude an atmosphere of well-to-do security, and the hotel staff is attentive, keeping a color-coded card file on guests' likes and dislikes. Yes, Elizabeth Taylor really does like pistachio nuts; Lord Snowden prefers French cigarettes.

Parts of the building date from 1912, and the posh hostelry with its lush landscaping has certainly seen its share of romance: Gable and Lombard, Yves Montand and Marilyn Monroe, even Paul Newman and Joanne Woodward—and they're married.

Then, there's the Polo Lounge, the fabled bar and dining room where Hollywood deals are made, often on pink or beige telephones plugged in table side. This is a good place to see and feel like part of the scene. Also, besides the patio dining area, there's the plush Coterie restaurant, and downstairs, the little coffee shop where famous faces breakfast quietly. Room rates run from $125 a day for a single to as much as $1,000 a day for a lavish bungalow or suite.

Off Sunset Boulevard, tucked away up Stone Canyon Road, the pink-

and-white **Hotel Bel Air,** 701 Stone Canyon Road (213-472-1211), seems far away from city strife. White swans glide along a small waterway lined with blossoming plants and orange trees. The Bel Air has been a favorite stopping place of such travelers as Richard Nixon, Helen Gurley Brown, Joyce Brothers, Barbra Streisand, and Dolly Parton. The late Princess Grace of Monaco was a favored guest. The hotel's quietly luxurious dining room overlooks a patio draped with purple bouganvillea. Room rates start at $140 for a single and go up to about $850 a day for a luxury suite.

The **Chateau Marmont,** 8221 Sunset Boulevard in West Hollywood (213-656-1010), is another peaceful retreat, even though it overlooks the day and night action of Sunset Strip. Dating back to 1925, the hotel has been declared a historical monument by the city of Los Angeles. The structure looks something like a French Norman castle, with what's been called a "Moorish" feeling. You can actually hear the birds singing in the small front garden, just steps from the Strip. The covered entrance walkway sports frescoes with painted nymphs and garlands.

Inside, there is little hustle and bustle, although the lobby's grand piano occasionally has been used to rehearse the not-so-sedate strains of *Hair.* After all, some 90 percent of the guests are from the entertainment world—stage, screen, and rock performers, writers and directors.

In the old days, recluses Greta Garbo and Howard Hughes often stayed here, as did such stars as Franchot Tone and Boris Karloff. The hotel has no bar or restaurant, just room service, but guests don't seem to care; Robert De Niro once stayed more than a year, and Tony Randall is a perennial visitor. The Marmont also has the grim distinction of being the spot where John Belushi died of a drug overdose.

The hotel's atmosphere seems generally homey and genteel. Rates run from $80 a day for a room, to $250 and up for a roomy bungalow, with various suites available in between. The penthouse suites goes for $275 a night, which is not bad for a comfortable, spacious apartment with a wood burning fireplace and a terrace with a panoramic view of the city and the Hollywood Hills.

And, for $100, true Chateau Marmont fans can buy a satin baseball jacket with the hotel's portrait embroidered in gold thread on the back.

The ornate facade of the **Beverly Wilshire,** 9500 Wilshire Boulevard, Beverly Hills (213-275-4282), faces the famed Rodeo Drive area. The hotel is actually two buildings divided by a specially constructed street, El Camino Real, made of cobblestones brought from Edinburgh castle. It was at the Beverly Wilshire that Japanese Emperor Hirohito was once greeted by traditional mariachi music.

Royalty are regular visitors. Prince Charles of England, plus the king of Spain and a plethora of Scandinavian princesses have stayed here, along with guests as varied as Louis Jourdan and Rudolph Nureyev.

There are several bars and restaurants with different international decors and flavors. Hernando's Hideaway is a hot spot for celebrities on Friday nights, as is the Pink Turtle coffee shop at breakfast time. Room rates go from $143 for a single to $1,000 a day for a luxury townhouse.

The **Century Plaza,** 2025 Avenue of the Stars, Century City (213-277-2000), is a sleek highrise just across from the ABC Entertainment Center. It's often hosted United States presidents: Johnson, Nixon, Carter, Ford, and Reagan. Hollywood and international royalty like Frank Sinatra and King Hussein are among the scores of celebrities who've stayed here. There are several restaurants and bars, but the lobby court bar is the best place for watching celebs and execs mingle. Room rates run from $110 for a single to $875 for the presidential suite.

Next door to Universal Studios in Universal City, the high-rise **Sheraton-Universal,** 333 Universal Terrace Parkway (818-980-1212), attracts a number of stars in town for tapings or meetings. The hotel exterior is also sometimes seen in commercials and television shows. The Sheraton has a full complement of restaurants and bars. Room rents run from $85 a day for a single to $125 for a luxury double.

The **Ambassador Hotel,** 3400 Wilshire Boulevard, Los Angeles (213-387-7011), is now surrounded by modern structures, but when it opened in 1921, the hotel was considered "out in the sticks" by hired help who had to walk seven blocks to work from the end of the nearest streetcar line.

It was here, among the twenty-three acres of greenery, that actress Pola Negri walked her pet panther—when it was very, very young, of course. Back in the Roaring Twenties, John Barrymore kept a monkey called Clementine in his bungalow for years. At one time, the Ambassador even had a mini-zoo.

The hotel has been a temporary home to Gloria Swanson, Al Jolson, Howard Hughes (he certainly seems to have gotten around), Omar Sharif, and John Wayne, among scores of others. Publishing magnate William Randolph Hearst and Marion Davies once rented half a floor. J. Edgar Hoover used to stay for a month every year.

The Ambassador is steeped in memories, both happy and not so happy. Senator Robert Kennedy was assassinated at the Ambassador. Between 1930 and 1943, the Academy Awards presentations were held at the hotel six times. Back in 1921, the famed Coconut Grove was Los Angeles's first real nightclub. Mary Pickford once picked up an Oscar there; stars like

Joan Crawford, Loretta Young, and Barbara Stanwyck used to join in the club's dance contests.

Nowadays, the lobby still has a mood of old-time opulence—a bubbling fountain, plenty of gilt, mirrors, and chandeliers. The sleek Palm Bar is a pleasant contrast and a comfortable spot for a quiet drink. Room rates start at $75 for a single and go up to $175 for the honeymoon suite, which comes with its very own pool. Quite a romantic bargain.

7

TAKING HOLLYWOOD HOME: MEMORABILIA

For thousands of years, pilgrims have returned from visits to shrines with relics and souvenirs and tales of what they have seen with their own eyes. To the devout movie fan, Hollywood is a Mecca in the California desert, the place where dreams are made. The booming market in Hollywood memorabilia proves that relics are as important today as in the past, and much more plentiful.

People want things that have been touched by the Hollywood magic for different reasons. Some are serious collectors or scholars interested in a particular genre or period. Some are investors, some are outright fans of a star or director, but most think it would be nice to see, touch, or own a small part of the magic. Hollywood has stores, to satisfy all these pilgrims.

The most collected Hollywoodiana are stills and posters. Stills are usually eight-by-ten glossies in either color or black and white; they can be posed portraits of an actor or actress, photos taken on the set, or even blowups from a single movie frame. The standard poster is called a 1-sheet and measures twenty-seven by forty-one inches. Larger displays are usually multiples of this and are called 2-sheets, 3-sheets, and 6-sheets. Lobby cards are smaller, indoor displays, almost always eleven by fourteen inches.

While photos and posters are the most plentiful collectibles, there are many other pieces sought after and available. Animation cels are the clear acetate sheets on which cartoon characters are drawn and colored; the cel is placed over a painted background and photographed so the character appears to be part of the scene. Other popular collectibles include autographs, scripts, costumes and costume sketches, props, books, magazines, and promotional tie-ins like toys, dolls, and other knickknacks. If it's been touched by Hollywood, someone wants a piece of the sparkle.

Another World Books & Comics, 1615 Colorado Boulevard, Eagle Rock (213-257-7757), stocks movie fan magazines and comic books, but you know what the main emphasis is when you pass through original prop doors from the sets of Buck Rogers and Flash Gordon serials. Stills and 1-sheets from science-fiction classics like *Forbidden Planet, The Day the Earth Stood Still, 2001,* and *Star Wars* abound, as do sci-fi magazines and film novelizations. Recent (post-1970) non-science-fiction pieces are also in good supply, ranging from Bruce Lee to Barbra Streisand. Most 1-sheets range from $5.00 to $10.00; eight-by-ten black-and-white stills are $2.50, while color runs $5.00. Phone and mail orders are welcome. The shop is open 10:00 A.M.–6:00 P.M. Tuesday through Saturday.

Animation fans can add to their collections, take a tax deduction, and benefit UNESCO at the same time. The world's largest sale of animation art is conducted every summer in Los Angeles by the **Association Internationale de Film d'Animation (ASIFA),** the international animation society, 1258 North Highland Avenue, Suite 102, Hollywood 90028. There are always some older items of interest offered, but most artworks are from current productions; prices range from $5 to $250. The location and date of the sale change yearly; contact ASIFA at 213-466-0341. Advice on the care and preservation of animation art is available by mail only.

Back Lot Books & Movie Items, 7278 A Sunset Boulevard, Hollywood (213-876-6070), is only open one day a week (Saturday from 10:00 A.M. to 5:00 P.M.) and can be hard to find (look for two garages at the rear of the lot), but the search is rewarding. Back Lot features over 5,000 1-sheets, ranging from $7.50 to $100.00; 1.5 million stills cataloged by title and actor ($1.50 to $3.00); and some fan magazines and sound track albums, but here uniqueness and price are the real drawing cards. Autographed eleven-by-fourteen stills of Marion Davies, Douglas Fairbanks, Sr., and Harold Lloyd fetch just $135.00. The same tariff can buy an autographed photo of all four Monkees. Other autographs can be had for as little as $5.00. An unusual selection of Polish posters for American films feature more art than Hollywood ballyhoo: *The Misfits* poster shows a stylized

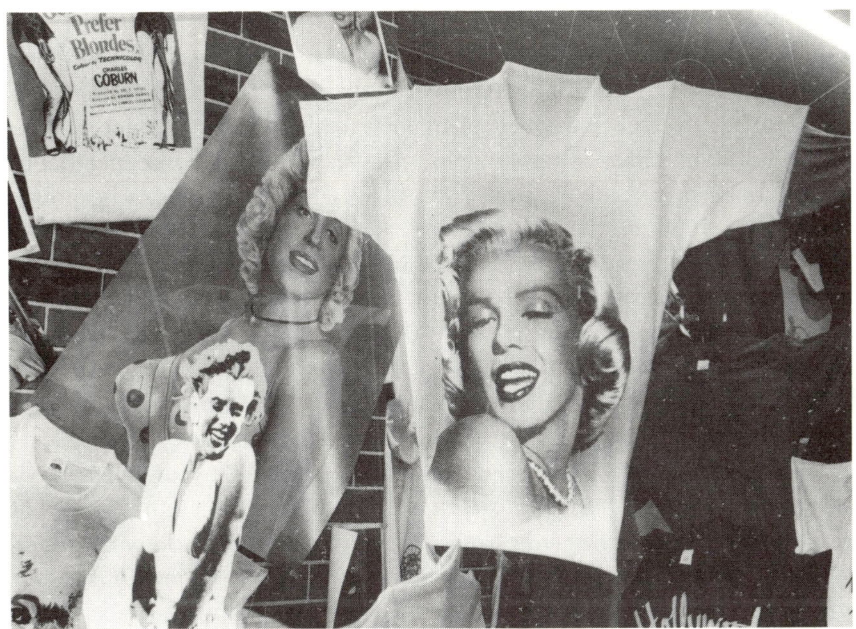

Good-bye—good buy—Norma Jean. . . .

horse and *no* Marilyn Monroe, while only her eyes are seen in one for *Niagara.* There's also a section of original, handsome forty-by-sixty lobby displays, popular in the 1940s and 1950s, which includes anything from Bela Lugosi to Ingrid Bergman as interpreted by sign artists ($95.00 to $135.00). A great place to browse.

Posters and lobby cards are the main focus at **Bijou,** in the Century City Shopping Center, 10250 Santa Monica Boulevard (213-277-0637), with examples dating back to the first talkies of the late 1920s. Original *Casablanca, King Kong, Gone With the Wind,* and *Wizard of Oz* 1-sheets, gallery framed, can go for as much as $1,600.00, but recent releases and reprints of the classics can be had for as little as $4.50 for lobby cards and $12.00 for 1-sheets. Bijou has over 20,000 pieces on file, including pre-1967 Disney animation cels and Broadway show posters. A separate department sells videotapes of classic films ($44.00 to $69.00) and, around Halloween, you can pick up an authentic Hollywood makeup kit for under $10.00. Bijou is open 10:00 A.M.–9:30 P.M. Monday through Friday, 11:00 A.M.–6:00 P.M. Sunday.

The specialties at **Book City Collectibles,** 6631 Hollywood Boulevard, Hollywood (213-466-0120), are photos, many autographed, and Beatle memorabilia. Candid stills and portraits of stars begin at $2.50 for eight-by-ten black and whites and $5.00 for color, but for a real one-of-a-kind bit of Hollywood how about Shirley Temple's *first* autographed photo, an eleven-by-fourteen sepia-toned print inscribed to her teacher, priced at $2,500.00. The thirteen-year-old Judy Garland remembered her dance instructor with a signed eight-by-ten black and white; it can be yours, framed, for $500.00. Marilyn Monroe, all five Marx Brothers, and James Dean are also represented, and a guarantee and letter of authenticity are included with each autograph. Beatle fans can pick up a free listing of almost 300 items. The shop is open 10:00 A.M.–10:00 P.M. Monday–Saturday, 10:00 A.M.–8:00 P.M. Sunday.

Burt Reynolds, Gene Hackman, and Diane Keaton are among the browsers and buyers at **Eddie Brandt's Saturday Matinee,** 6310 Colfax Avenue, North Hollywood (818-506-4242). Shoppers are drawn by a selection of more than 20,000 poster titles and one million stills, as well as press kits and movie magazines dating back to the 1920s. Lobby cards and 1-sheets range from $10 up; top-end items include an original *Citizen Kane* card ($350) and one from *The Wizard of Oz* ($750). Brandt stocks about 200 autographed pieces; try an eleven-by-fourteen Buster Keaton portrait from the 1920s at $250. Stills are filed by actor and title and begin at $2 for black and white and $5 for color. Brandt's is open 8:30 A.M.–5:00 P.M. Wednesday through Saturday.

The **Burbank Book Castle & Movie World Annex,** 200–212 North Golden Mall, Burbank (818-845-1563), is a supermarket of print memorabilia, with over one million items, many surprisingly low priced: lobby cards sell from 50¢; 1-sheets are as low as $1 but average $6 to $7. Books, fan magazines, and early *TV Guide* and *Life* magazines with movie-star covers are plentiful, and the Castle even stocks casting directories from the twenties and thirties. Original movie and television scripts, most of which carry production notes and call sheets, were actually used by cast and crew members during filming. An old "Gunsmoke" or "Superman" script will fetch $10 to $30, and many recent feature scripts are available starting at $1. The shop is open 10:00 A.M.–8:00 P.M. Monday through Thursday, 10:00 A.M.–9:00 P.M. Friday, and 10:00 A.M.–6:00 P.M. Saturday and Sunday.

At **Chapman's Picture Palace,** 1757 North Las Palmas Avenue, Hollywood (213-467-1730), portraits of stars are the feature. Eight-by-eleven black and whites are priced below $3.00; color eleven by fourteens range from $7.50 to $15.00. Most of the subjects are from the golden age of Hollywood; Garland, Monroe, and Bette Davis are examples. Occasionally, Chapman's will organize a showing of the works of well-known portrait photographers. Chapman's is open noon–6:00 P.M. Wednesday through Saturday.

Chic A Boom, 6905 Melrose Avenue, Hollywood (213-931-7441), has a *huge* collection; in addition to the store, two warehouses and a garage hold thirty-three categories of memorabilia. In the movie and television department, the store alone stocks over 10,000 posters and 6,000 lobby cards, but it's the entertainment-oriented collectibles that stand out. A mint condition Howdy Doody puppet in the original box is $110; there are hundreds of similar dolls, toys, promotional pieces, and even board games from the early days of television. The Beatles, James Bond, and Batman rate special collections, as do movie sheet music, magazines, and *New York Times* rotogravures from 1911 to 1918. Chic A Boom, a place a movie fan could get lost in, is open 11:30 A.M.–7:00 P.M. Monday through Saturday.

In addition to eight-by-ten stills (50¢ to $20) and posters and lobby cards ($1 to $30), **Chuck & Rita's,** 5515 Lankershim Boulevard in North Hollywood (818-761-2201), features some props and devices actually used in moviemaking. The four-foot-long special-effects tank used in filming *Captain Nemo,* starring Chuck Connors, is on display and for sale for $1,000. Futuristic firearms from *Voyage to the Bottom of the Sea* and "Lost in Space" sell for $50 and up; a dagger from *Ben Hur* is negotiable. There are more than 200 autographed items from $15; Elvis fetches $185, Mae West $90, and Gloria Swanson $100. There's also a selection of old cereal-box premiums from radio and early television. The shop is open 11:00 A.M.–5:00 P.M. Monday through Saturday.

Collectors Book Store, 6763 Hollywood Boulevard, Hollywood (213-467-3296), is probably the largest and most complete movie-oriented store in Hollywood, stocking 75,000 posters, 2 million catalogued stills, and 50,000 film and television scripts. Some rarities include a series of Edison movie posters dating to 1912 from William Randolph Hearst's collection ($100.00 to $125.00) and a never-released *Revenge of the Jedi* 1-sheet (the *Star Wars* sequel was retitled *Return of the Jedi*). Collectors buy many pieces from Hollywood estates, especially stars' autographs. Greta Garbo's signature, rare in its own right, is worth $1,200.00 when it's on the bottom of a contract. The store features a monthly mail auction of such one-of-a-kind items as scripts, animation cels, posters, magazines, and original art; the catalog is $1.50. This is a store where there's always something new on sale and someone to collect it. How many of your frieds own a "Battlestar Galactica" warrior's costume ($125)? The store is open 11:00

A.M.–5:00 P.M. Tuesday through Friday and 10:00 A.M.–5:30 P.M. Saturday.

The great leading ladies are remembered at **Dorothy's Surrender,** in the French Market Place at 7985 Santa Monica Boulevard, West Hollywood (213-650-4111). There's a large collection of out-of-print recordings of the warblings of Marilyn Monroe, Marlene Dietrich, and Judy Garland, as well as many magazine covers featuring leading ladies of the twenties, thirties, and forties. There's also some real esoterica. Try Joan Crawford's ankle-strap shoes on for size—just $300 with matching handbag. Judy Garland's dresses begin at $1,000. One-of-a-kind autographed candid snapshots of such luminaries as Tallulah Bankhead, Dietrich, and Mae West run $25 to $50. Also, original sixteen-by-twenty matted ads featuring movie-star endorsements—Mae West for Lux Soap?—are just $8 to $20. The shop is open noon–11:00 P.M. Sunday through Thursday and noon–midnight Friday and Saturday.

First and foremost a bookstore, **Larry Edmunds Cinema and Theater Book Shop,** 6658 Hollywood Boulevard, Hollywood (213-463-3273), stocks just about every book on films and plays that are in print and thousands that aren't, and film magazines from the 1920s on. But also in residence are over 100,000 1-sheets (from $10.00 to over $2,000.00), eight-by-ten stills (black and white, $2.00 to $5.00, and color, $5.95 and up), and autographed photos. Some notables: a fourteen-by-thirty-six *Gone With the Wind* lobby display ($750.00), a 6-sheet poster from the 1914 cliffhanger *Perils of Pauline* ($2,500.00), and a 1920s autographed photo of Charlie Chaplin ($600.00). If this is a bit out of range, autographed portraits of current stars, like Burt Reynolds, start around $5.00. The shop is open 10:00 A.M.–6:00 P.M. Monday through Saturday.

Fantasies Come True, 7408 Melrose Avenue, Los Angeles (213-655-2636), is the place for your class reunion if you're a Mickey Mouse Club alumnus. The only language spoken here is Disney. At any given time there are twenty-five to thirty original Disney cels on display, ranging from $18 to $350. Original black-and-white drawings of Jiminy Cricket, the Three Pigs, and Goofy average $50 to $60. A packet of rare 1931 Mickey Mouse sparklers commands $330; an example of the first velvet Mickey hand puppet, also from 1931, is $530. There are more than 5,000 toys, records, and knickknacks that are part of the lives of the famous rodent and his creator. And, of course, you can buy an original Mickey Mouse wristwatch; it keeps time and costs just $300. Fantasies Come True is open noon–4:00 P.M. Monday through Wednesday, Friday, and Saturday, or by appointment.

At the **Hollywood Book & Poster Co.,** 1706 North Las Palmas Avenue,

Hollywood (213-465-8764), the stock includes more than 500,000 posters and lobby cards. Posters from recent releases and reproductions of classic posters range from $1 to $4; rarer originals range up to $200. Most black-and-white stills are $2; color collectors can have 35mm slides for just 25¢. Sixteen and 35mm movie trailers (the "coming attractions") are thirty seconds to three minutes long, with sound, and cost $1 to $10. Matted autographed stills range from Clint Eastwood and Burt Reynolds to James Cagney and Bette Davis, and average $5 to $50. Separate departments deal in movie sheet music and scripts from $1 to $25. The shop is open 11 A.M.–7:00 P.M. Monday through Thursday.

While general interest books are the mainstay at the **Hollywood Book Store & Search Service,** 1654 North Cherokee Avenue, Hollywood (213-464-4164), there is a large film section. The attraction to collectors, though, is the file of 40,000 pictures of stars, displayed on proof sheets—*lots* of proof sheets. Marilyn Monroe takes up thirty-nine sheets, Errol Flynn fourteen. Once you choose, an eight-by-ten black-and-white print will cost $3.65; some color is available at $5.00 and up. Prints can be made in other sizes for an additional charge. The shop is open noon–5:00 P.M. Monday through Friday, noon–3:00 P.M. Saturday.

At **Hollywood Movie Posters,** 6727½ Hollywood Boulevard, Hollywood (213-463-1792), you'll find a straightforward poster-and-still store, with a slight emphasis on science fiction and horror. The stock of 6,000 eight-by-ten stills is divided evenly between black and white ($2.50 to $3.50) and color ($4.50). The 1-sheets average $25.00, but classics can range higher: original posters from *War of the Worlds, Forbidden Planet,* and *The Day the Earth Stood Still* are $300.00 to $400.00; *Psycho* brings $100.00. Unframed autographed pictures, including Alan Ladd, Bette Davis, Clark Gable, and Dorothy Lamour, run between $35.00 and $100.00. They'll mount your poster on protective acid-free linen backing for about $50.00. The store is open 11:00 A.M.–6:00 P.M. Monday through Saturday.

Professionals from the decorators of the MGM Grand Hotel in Las Vegas to movie set designers shop for authentic period pieces at the **Hollywood Poster Exchange,** 965 North La Cienega Boulevard, Los Angeles (213-657-2461), as have stars such as Suzanne Pleshette, Bob Hope, and the late Groucho Marx. They, and the public, can select from over 20,000 original stills ($2.50 to $10.00), lobby cards ($7.00 to $8.00), and 1-sheets ($30.00 average). Above-average items include a *Gone With the Wind* 1-sheet ($1,800.00), Mae West in *Klondike Annie* ($800.00), and a first-release *Pinocchio* ($900.00). There are also unique pieces displayed: a dressmaker's mannequin used to costume Joan Crawford is $300.00, and a rubber mask used in the epic feature *Abbott and Costello Meet Dr. Jekyll and Mr. Hyde* costs $175.00. Autographed 1-sheets include James Stewart in *Destry Rides Again,* Jane Russell in *The Outlaw,* and Groucho in *A Night at the Opera.* These are available framed, start at $100.00, and can be seen 11:00 A.M.–5:00 P.M. Tuesday through Saturday.

Television and movie researchers make good use of **Marlow's Bookshop,** 6609 Hollywood Boulevard, Hollywood (213-465-8295). Period fan magazines, books, and stills ($2.50) are prime items, as are files, indexed by name, covering entire careers of stars. There's also a large collection of used books on film, at discount prices. Marlow's hours are 10:00 A.M.–6:00 P.M. Monday through Thursday, 10:00 A.M.–9:00 P.M. Friday and Saturday, and noon–6:00 P.M. Sunday.

Although the **Movie Poster Warehouse,** 1550 Westwood Boulevard, West Los Angeles (213-470-3050), has some eight-by-ten black-and-white and color portraits dating to the 1930s, as the name implies, posters are the main stock in trade here. The emphasis is on Hitchcock films, from *Notorious* and *Strangers on a Train* through *Psycho* and *The Birds* ($50 to $500). There's also a line of foreign, particularly French, 1-sheets advertising American films by such directors as Preston Sturges and John Ford. And the Warehouse offers a real bargain: top-quality, full-color photo reproductions of 1-sheets from classics like *Casablanca* go for $12.50, a fraction of the costs of the originals. Open 11:00 A.M.–5:00 P.M. Wednesday through Saturday.

The word is **Nickelodeon,** 13826 Ventura Boulevard in Sherman Oaks (818-981-5325), is eclectic. First, of course, are the posters and lobby cards, Monroe in *Some Like It Hot* and Cooper in *The Plainsman* merit mention ($85.00 to $100.00). Then there are the photos, several hundred stills that include a color sixteen-by-twenty print of the famous 1949 MGM group photo from *Life* magazine for $45.00. Autographs? Well, Pickford and Fairbanks of course ($75.00), but Burt Reynolds's signed Actors' Equity card? Relics? You can get one of the seven plaster King Kongs used in the Dino De Laurentiis movie for $300.00. But the main emphasis here is on Walt Disney. Not only are there hundreds of items like Mickey Mouse watches, dishes, radios, and toys (the rare Mickey Mouse handcar is available for $650.00), but owner Ed Levin runs a club for Disney collectors. Membership is $17.50 a year; there is a bimonthly newsletter and an annual three-day convention in Anaheim. The store is open 11:00 A.M.–5:00 P.M. Wednesday through Friday and noon–5:00 P.M. Saturday.

There are two reasons to visit the **Retake Room,** 3953 Laurel Grove Avenue, Studio City (818-508-7762): designer clothes at super prices, and clothes that have been used in your favorite movies and television shows. Some costumes, but mostly streetwear, are consigned by the major studios to this store after they've been worn by actors, frequently only once or twice, in a produciton. You can find fashions by Givenchy, Blass, and Carol Little, as well as clothing used in *Sophie's Choice, On Golden Pond,* and *Dead Men Don't Wear Plaid,* to name a few. At the store's low prices, alterations are the customer's responsibility, but if you happen to be built like Julie Newmar, her Renaissance dress from *Hysterical* should only need a tuck here and there. The Retake Room is open 10:30 A.M.–5:30 P.M. Monday through Friday and 10:00 A.M.–6:00 P.M. Saturday.

When Warner Bros. wanted their file of signed documents appraised they called **The Scriptorium,** 427 North Canon Drive, Beverly Hills (213-275-6060). Autographs are the specialty here, and not just on scraps of paper, but letters, documents, and uncommon photos signed by the famous. A set of letters from Grace Kelly to an ex-lover describing life in the palace in Monaco is priced at $1,500; one of Judy Garland's MGM contracts is worth between $120 and $250. Autographed photos begin at $10 and run as high as $1,500 for a very rare example of Greta Garbo's signature. Each of the 2,000 to 5,000 items here is authenticated and fully guaranteed. The Scriptorium is open 11:00 A.M.–6:00 P.M. Tuesday through Saturday.

If science fiction is your mania or if you just like a bargain, make the trip to Long Beach and check out **The Tape and Record Room,** 201 East Broadway (213-432-7602). Sci-fi and fifties horror-movie posters start at $1 and most are under $10. A Presley *Frankie and Johnnie* 3-sheet is only $35, as is a 3-sheet from *The Empire Strikes Back.* An original fourteen-by-thirty-six *Rebel Without a Cause* poster is $150, and a real bargain is a Disney *Alice in Wonderland* poster of the same size for just $100. Drop in 9:00 A.M.–6:00 P.M. Monday through Saturday.

You step into the art deco world of the twenties and thirties when you enter **Tiberio,** 458 North Robertson Boulevard, West Hollywood (213-659-5777). Here you can buy the art deco statue used in Joan Crawford's *The Women* for just $5,000. If that's too rich for your blood, you can choose among autographed photos of Crawford, Harlow, and Monroe—or even Crosby and Hope. Prices start at $35, but be prepared to spend $1,000 for Marilyn. Tiberio also supplies the movies as well as movie fans; the store furnished all of Faye Dunaway's period jewelry in *Chinatown* and a period magazine stand with 1938 magazines for *The Day of the Locust.* Tiberio is open 10:00 A.M.–6:00 P.M. Monday through Saturday.

In among the antique furniture and Western art at **Xanadu Galleries,**

212 North Orange Avenue, Glendale (818-244-0828), are some truly rare movie collectibles. Xanadu recently acquired the Wirtzel collection of first-generation eleven-by-fourteen star portraits from the twenties; over 175 prints of Fatty Arbuckle, Mable Normand, Pola Negri, and the other silent-screen greats can be had for $7,500. The oil painting of Joan Crawford that hung over the fireplace in *Mildred Pierce* is $1,500, but B movie 1-sheets start at just $3. There's good browsing through movie books and Chaplin, Pickford, and Fairbanks memorabilia. Xanadu is open 10:00 A.M.–6:00 P.M. Monday through Friday and 11:00 A.M.–4:00 P.M. Saturday.

Movieland Books

New Media Publishing Inc.
3530 Mound View Ave.
Studio City, CA 91604

$10.00 UNCLE #1

Epilogue

Hollywood has been called many things, from the dream factory to a trip through a sewer in a glass-bottom boat. Some people love it; some people hate it; a lot of folks, including a good many of us who live and work here, do both. Few people ignore it, though, any more than they could ignore King Kong if they woke up one morning and found him in their beds. Like the man who came to dinner, the movies, and the movies' kid brother, television, crept surreptitiously into our lives, slowly insinuating themselves so that they had become permanent residents of our psyches' households before we even knew they had dropped in.

One example, perhaps. A few years ago, before Henry Fonda died, I remember thinking that I had never paid much attention to him as an actor; if hard-pressed, I doubt that I could have named more than three or four of the movies he had made. Then, at an Academy Awards presentation, the Academy ran a compilation of clips from Fonda's films, and suddenly in those few minutes I was overwhelmed by images that I had carried around in the back of my head since I was a boy. Suddenly, there once again was Tom Joad, driving his hulk of a truck from the dust bowl to California, looking for a better life; there was the ex-con in *You Only Live Once,* carrying the body of his lover toward the shining light in the fog, hoping not just for a better life, but for a better world. There was Mr. Roberts, there was Wyatt Earp; there were *She Wore a Yellow Ribbon's* cavalry captain and *The Ox-Bow Incident's* lone anguished man.

I knew Henry Fonda better than I thought, it seemed. I had grown up with him, or, we had grown up together, he faster than I, since I had seen him age and grow through four decades of filmmaking in just a few years. In the end, I realized, I knew him—or at least the characters he played—as well as I knew many of my friends. And, knowing him in that way, I felt the loss when he died.

In a display case in the Motion Picture and Television Museum at the Universal Studios Tour are the fishing license and old, beat-up fisherman's hat Fonda wore in *On Golden Pond.* In a separate case across from them is the Oscar he won just before he died. The first time I saw it, I started remembering how much of my vision of the world I owed to Fonda and to the hundreds of other men and women who made the movies I remembered, from John Ford and Steven Spielberg to Laurel and Hardy and Bogart and Bacall. Together, they had helped make my world, and watching them in darkened theaters, I had made their worlds mine.

Maybe you're looking for someplace new and exotic to visit; Antarctica, maybe, or Yemen. Maybe Borneo's nice this time of year? Forget it. Come to Hollywood instead. You'll see locales from China to New York, the Old West to Frankenstein's village, Casablanca to the South Pacific to Wales. You'll get to visit Beaver's old neighborhood and see the house where E.T. lived. You can go around the world—heck, around the universe—in eighty minutes instead of eighty days, with enough feelings of *deja vu* thrown in to last a lifetime. And if anyone mentions that some of the locales and the characters and creatures who inhabited them aren't real, just remind them that the feelings they engendered when you saw them on the screen certainly were—and that's the strongest reality of all.

Bibliography

Balio, Tino. *United Artists.* Madison: University of Wisconsin Press, 1976.

Baxter, John. *Sixty Years of Hollywood.* South Brunswick, N.J.: A.S. Barnes & Co., 1973.

Birnbaum, Stephen. *Steve Birnbaum's Guide to Disneyland.* New York: Houghton Mifflin Co. & Diversion Communication, 1982.

Blum, Daniel. *A Pictorial History of the Silver Screen.* New York: Grosset and Dunlap, 1953.

Brownlow, Kevin, and Kobal, John. *Hollywood: The Pioneers.* New York: Alfred A. Knopf, 1979.

Cini, Zelda, and Crane, Bob. *Hollywood: Land and Legend.* Westport, Ct.: Arlington House, 1980.

Clarke, Charles G. *Early Film Making in Los Angeles.* Los Angeles: Dawson's Book Shop, 1976.

Daggett, Dennis Lee. *The House That Ince Built.* Glendale, Ca.: Great Western Publishing Co., 1980.

Edmonds, I. G., and Mimura, Reiko. *Paramount Pictures and the People Who Made Them.* San Diego/New York: A. S. Barnes & Co., 1980.

Fernett, Gene. *Poverty Row.* Satellite Beach, Fla.: Coral Reef Publications, 1973.

Gebhard, David, and Winter, Robert. *A Guide to Architecture in Los Angeles & Southern California.* Santa Barbara/Salt Lake City: Peregrine Smith, 1977.

Grenier, Judson A. *A Guide to Historic Places in Los Angeles County.* Dubuque: Kendall/Hunt, 1978.

Halliwell, Leslie. *Mountain of Dreams.* New York: Stonehill Publishing Co., 1976.

Higham, Charles. *Warner Brothers.* New York: Charles Scribner's Sons, 1975.

Hurst, Richard Maurice. *Republic Studios: Between Poverty Row and the Majors.* Metuchen, N.J./London: The Scarecrow Press, 1979.

Lamparski, Richard. *Lamparski's Hidden Hollywood.* New York: Simon and Schuster, 1981.

Larkin, Rochelle. *Hail, Columbia.* New Rochelle, N.Y.: Arlington House, 1975.

Lockwood, Charles. *Dream Palaces.* New York: The Viking Press, 1982.

Lunsford, James W. *Looking at Santa Monica.* Santa Monica, CA: James W. Lunsford, 1983.

Manes, Stephan. *Sixty Years of Studio.* Los Angeles: Stephan Manes, 1971.

Marill, Alvin H. *Samuel Goldwyn Presents.* South Brunswick, N.J.: A. S. Barnes & Co., 1976.

Marinacci, Barbara, and Marinacci, Rudy. *Take Sunset Boulevard.* San Rafael, CA: Presidio Press, 1980.

Murphy, Bill. *The Dolphin Guide to Los Angeles and Southern California.* New York: Doubleday & Co., 1962.

Parish, James Robert. *The RKO Gals.* New Rochelle, N.Y.: Arlington House, 1974.

Schessler, Kenneth. *This Is Hollywood.* Los Angeles: Gresha Publications, 1978.

Slide, Anthony. *The Kindergarden of the Movies.* Metuchen, N.J./London: The Scarecrow Press, 1980.

Stanley, Robert. *The Celluloid Empire.* New York: Hastings House, 1978.

Torrence, Bruce T. *Hollywood: The First Hundred Years.* New York: New York Zoetrope, 1982.

Wanamaker, Marc. *Encyclopedia of the Movie Studios.* Unpublished.

INDEX

ABC Television Centre	213/557-7777
TBS (Warner + Columbia)	818/954-1744 -1008
NBC Studios	213/840-3572
Paramount	213/468-5000
20th Century Fox	213/277-2211
Hollywood on Location	213/659-9165